I0063225

"*Navigating Leadership* colorfully outlines leadership tenets using a compact, step-by-step approach packed with wit and wisdom. It is fresh, readable, and energizes and inspires individuals to realize their own leadership potential. Anyone who possesses the desire and will to lead will find this book easy to read but impossible to forget!"

-- Linda K. Marion, M.Ed.
Instructional Leadership, Author of Powerteaching for the 21st
Century: A Management Guide to Integrating Curriculum

"Admiral Hall sent me the manuscript and asked me to scan it and possibly make a comment. I started my scan and it turned into a thorough reading from cover to cover. I have been involved in Leadership my entire life as a Marine (1971-2008) and then as the Leadership Chair at the US Naval Academy (2011-2017). I strongly believe Leadership is a combination of Character and Competence. Admiral Hall's life journey covers the key elements of a Leader in a fun, interesting form. The importance of Passion, Accountability, Commitment, Empowerment, Trust and Humility are driven home. All built on a foundation of Selfless Service and the power of Appreciation (Thank yous motivate!). I believe his story will reinforce what you already know/believe and may teach you a new thing or two about yourself. Recommend reading with a yellow highlighter and pen."

-- John Sattler
Lieutenant General, United States Marine Corps
Former Chair of Leadership at the United States Naval Academy

"Admiral Garry Hall is an accomplished Naval Aviator, strategic leader and the consummate storyteller. In Navigating Leadership, Admiral

Hall shares his leadership philosophy through principles and personal stories of challenge, triumph, laughter and failure and highlights the fact that leadership is a deeply personal and lifelong journey of growth and development. Navigating Leadership is an entertaining and inspirational read for leaders young and old!

-- Stephen Trainor, Ph.D. Captain, USN (ret.)
Faculty Lead, Walton College Executive Education
Founder and Chief Explorer, Expeditionary Leadership

"While there are many books on leadership on the market, Admiral Garry Hall's Navigating Leadership is unique and is a lively read. Most books on leadership focus on people occupying the C-suite, senior military officers, or political figures. Hall's book is more broadly focused and provides insightful leadership lessons that everyone can apply. If there is a book that every person who wants to motivate people to do what they don't think they can do, or don't want to do, this is the one."

-- George Galdorisi
Captain, United States Navy (retired),
Best Selling Author of Fiction and Non-Fiction

"Navigating Leadership" offers readers an entertaining and thought-provoking autobiographical sketch of Admiral Garry Hall's journey of leadership development and personal growth. By skillfully weaving leadership lessons from his experiences and challenges, he offers exercises and strategies to improve your own leadership skills. With its engaging narrative, wit, and practical advice, Navigating Leadership is a must-read for anyone looking to become a better leader and a more effective individual."

-- Sanford B. Ehrlich, Ph.D.

"Admiral Hall has served the nation with honor and distinction. His military career is full of remarkable experiences. He has a brilliant ability to draw valuable insights from them and share in a way that is easily understood. He is also a gifted speaker who can enthrall an audience."

-- Dan LaBert
President, FWA Consultants

"Navigating Leadership" by Admiral Garry Hall is an absolute must for all students of leadership ... a compelling, brilliant, and inspirational offering of experience from a sharp-witted, engaging, and truly experienced leader who has led from the decks of Naval ships and hallways within the Whitehouse while serving as a National Advisor to the President. Garry possess a sharp-witted gift of humor, providing a 'spark' for others, by demonstrating that leaders can be real people, in touch with what really matters, connecting on the human-level in a personal, disarming and comfortable manner that enables others to relax, share, and trust this leader's judgement, knowing that they will be protected, and led in a direction of hope, opportunity, and success. Read and enjoy, as you reflect on the golden nuggets of lessons learned, and the pearls of wisdom from the rich experience base of this uniquely qualified, and successful national leader.

-- Scott Hartwick
President and CEO
United Space Alliance

CONFIDENCE
Whether it be standing a watch, leading the troops, or responding to a National Security incident, as a young leader, confidence is critical to success.

I was fortunate enough to be a Junior Officer under Admiral Hall's leadership. He embraced me and my young family, he developed my confidence and taught me how to develop confidence in others. As I work to grow my company, I rely on the humility and wisdom that Admiral Hall developed in me.

In my humble opinion, the ability to cultivate confidence distinguishes good leaders from great leaders and Admiral Hall is, without question, one of the greats!

Thank you, Garry, for all that you have done for me.

-- Drew Kellogg
CEO Oath Pizza

Rear Admiral Garry Hall, USN Ret, has written an absolute "Must Read" primer to anyone seeking a leadership style that is unmatched. Written with Admiral Hall's entire story, start to finish on how to lead effectively under all circumstances and hurdles.

-- Captain John "Tuck" Shattuck
Former Naval Aviator, 30+ Years Commercial Airline Pilot and CEO,
President, Owner of a successful S-Corp, USA

Published by Mediacasters Publishing House

ISBN: 979-8-9883028-1-0

NAVIGATING
LEADERSHIP

Making a PACT with Excellence

Published by Mediacasters Publishing House

THE
MEDIA-
CASTERS

DEDICATION

With love and affection, I dedicate this book to the love of my life, Darlene Allred Hall. It was love at first sight for me and we married six months to the date of meeting. Forty-five plus years have passed and she makes me smile and laugh every day. Texas tough, Texas savvy, and Texas loving, she has taught me unconditional love, made a home for our family in temporary quarters, military housing, and other houses and quarters around the globe. She sees wonder and beauty in all things. She is my secret weapon for success and everyone who has met her knows I'm the luckiest man around. Life with Darlene is always an adventure.

To my children Garry, Courtney, and Sam who did not volunteer for the Navy life but willingly traveled along with upbeat attitudes and travelers' spirits. Life with multiple homeports, neighborhoods, schools, and friends have made them the closest allies. They have grown into wonderful, compassionate adults and astute citizens of the world.

To every Sailor and Marine I served alongside and had the honor to lead, they are the best that this country has to offer. They took an oath and served with honor and made the world a better place during chaotic times of peace, world tensions, and war. They are servant leaders.

TABLE OF CONTENTS

FOREWORDS

One day the doorbell of my rectory rang. When I answered it, I met a tall, distinguished gentleman. He said with a big smile, "My name is Garry Hall. I just retired from the Navy and I am between jobs. I'll mow the lawn or file your files." With his characteristic laugh and generous smile, he continued, "Anything or else my wife is going to divorce me. She's sick of my hanging around the house."

Garry was quite literally an answer to a prayer. At the time, I was pastor of Saint Peter's on Capitol Hill, a large urban parish in Washington, DC. The House of Representatives is within its boundaries. With Hill staffers, lobbyists and politicians, it is a busy place. I was also serving on the Archbishop's staff running pastoral ministry in the Archdiocese of Washington. I felt strongly that God wanted me to do a special project in the parish; but with my current workload, I didn't know how I was going to get it done. I told God that I'd be happy to start the project, but I needed God to send me some help. He sent Garry the very next day.

I asked Garry to help me organize small group Bible studies throughout the parish. Before I knew it, we had more than 100 people engaged in lively discussions in people's homes. He had also created a manual so that anyone could take over the project at any time and help it thrive. I knew then that God had sent me the great leader. Over time Garry, his beautiful wife Darlene and I developed a wonderful friendship. While I spoke to him about spiritual growth, he helped me learn to be a better leader.

Several years later, I received an unexpected call from an official of the Vatican informing me that Pope Francis had named me to be the Bishop of Springfield, Massachusetts. Suddenly, I went from overseeing one large parish in suburban Maryland to overseeing the Catholic Church in western Massachusetts with 79 parishes and many schools. It was then that I entered into more intentional discussions with Garry Hall about leadership. The baby Bishop was having a Master Class with the Admiral!

Navigating Leadership is a wonderful book. I speak from firsthand experience of the valuable insights that Admiral Garry Hall has given me and now offers to us all. Everyone, regardless of your experience or of the heights to which you have risen, can always be a better leader. Garry's wisdom is both practical and easily implemented to enhance one's style of leadership in any given situation. Above all, Admiral Hall will keep you laughing.

-- Most Reverend William D. Byrne
Bishop of Springfield, Massachusetts

A few years ago, I saw a sign in a general store that read: "Irish diplomacy is the ability to tell a man to go to hell so that he looks forward to the trip." I remember thinking at the time how perfectly that phrase also captured the challenge of leadership in the crucible of combat or under other very stressful and difficult situations. Great leaders are those who can inspire ordinary men and women to do extraordinary things regardless of the circumstances. Indeed, they are servant leaders we will enthusiastically follow to accomplish the mission even at significant personal risk. Garry Hall is such a leader.

I met Garry almost 20 years ago as a fellow student at a course conducted for newly selected flag officers from the United States Army, Navy, Air Force, Marines and Coast Guard. It was immediately obvious to those of us in the class that that he had an unconventional style and a gift for making people laugh. In fact, during the course, his often-irreverent sense of humor was a constant. In a profession known for conformity, Garry Hall clearly had his own style. I was immediately drawn to Garry as a "liberty partner" during the travel related to the course. But as our relationship and valued friendship developed over time, I learned that there was much more to Garry than natural charisma. He is also a man of rock-solid character, exceptional operational competence, and a passion for taking care of people. His sense of humor is accompanied by exemplary professionalism. Throughout his distinguished Navy career, Garry Hall earned the complete trust, admiration, and affection of those in his charge while leading from the front. He has a well-earned reputation for building teams with esprit de corps and a commitment to excellence. There is a legion of Marines and Sailors, Joe Dunford included, who would follow the Admiral anywhere!

'Navigating Leadership' is Admiral Garry Hall's gift to those who aspire to follow in his footsteps. He shares his personal leadership journey with candor, humility, and keen insight. Aspiring leaders from all walks of life will be better prepared to navigate their own path if they fill their seabags with his wisdom, lessons learned, and leadership principles.

-- Joe Dunford
General, USMC (Ret.), 19th Chairman, Joint Chiefs of Staff

A NOTE FROM THE ADMIRAL

As you dive into the pages of this text, I hope to convey the message that if you desire to lead, then continual leadership development is an essential element for success. By mastering the fundamentals of leadership, you can inspire and guide your team towards achieving their goals and your organization towards greater heights.

Leadership is not just about giving orders or managing people, it's about understanding the needs and motivations of individuals, and effectively communicating with them. It's about building trust, providing support, and creating an environment where everyone feels valued and heard.

Leadership skills are not innate; they can be learned and honed through practice and dedication. By investing in your leadership development, you can become a better leader and positively impact those around you.

So, let's embark on this journey together and explore the key principles of leadership. I assure you that the knowledge and insights you gain from these pages will serve as a solid foundation for your leadership journey and bring about positive change in your personal and professional life.

Garry

INTRODUCTION

President John F. Kennedy once said, "Conformity is the jailer of freedom and the enemy of growth." These eleven words summarize my unconventional ascension as I navigated the hallowed halls of The White House as a Navy admiral.

I am not your typical leader. In this book, you will have VIP access to a maverick who learned lessons the hard way. My hope is that with each page turned, you will understand what authentic leadership means to me.

I take pride in my disarming wit and penchant for magic tricks. I am a leader who pushed the envelope in combat and had "a come to Jesus moment" that left me rethinking everything. However, rising the ranks is not for the faint of heart. Leadership must be born with a purpose and supported with character. When hardship prevails, that sense of purpose and strength of character are the secret weapons that combat the enemy.

The ironic part about combating the enemy is that we find that we are often our biggest adversary. When we are not the biggest champion of our success, we find ourselves in places of self-sabotage or feeling like an imposter. So often, we find ourselves on a teeter-totter that swings from self-doubt to overconfidence. The truth is, the hardest person to lead is yourself.

My childhood was not extraordinary, and there is nothing wrong with being ordinary. However, from the moment I was born, I had an insatiable hunger to be something more than my socioeconomic birthright allowed.

Taming this unruly boy from Buffalo, New York, was an arduous task. I was a quirky, energetic kid who craved love and attention. I'd be scolded by teachers and punished by my parents. However, I knew I needed to make a difference in this world. At times, I felt tethered to an ordinary life, so I kept fighting to do more and be more. Although I was admonished by my early leaders (mother, father, teacher, and preacher), my untamed spirit could not be extinguished.

My ascension to my rank was anchored with perseverance, disarming commentaries, and an abiding dedication to God and Country. I was not destined to be a man cloaked in medals and accolades. My modest background was not the fertile land in which leaders flourish. In fact, it was sown by two dysfunctional parents who could not understand my energy. My road to leadership was provided by the rigors of naval life and the steadfast love of my wife, Darlene, who helped me cultivate the compassionate leader I am today.

We should look leadership square in the face and choose our individual paths. Your road may take you to commanding a fleet as the Admiral in the Navy. Or perhaps the trail you follow will lead to starting an innovative company, improving the culture at a Fortune 500, or leading a non-profit.

On the other hand, leadership can be as profound as committing to being a stay-at-home mother or venturing into a new world as a college student. We are all born to be excellent. The secret is harnessing your unique character and fortitude to overcome inevitable rough waters.

Bumps along the road are all too familiar to me. I am the self-proclaimed commander of *the major-bumps-in-the-road action plan*. On the surface, I have an unsurpassed resume. My pedigree includes rising in the ranks of the Navy and a prestigious tenure in The White House. However, the environment I was born into was not the conventional background for a leader. My dysfunctional life was an incubator of dissatisfaction and contempt. However, failure was not an option.

If you're reading this book, you're a leader, be it emerging or experienced. You may be in the military, work in a corporate landscape, or have just graduated from college. I hope this book and my journey will make you rethink leadership and thus improve your life in all areas and aspects.

Leadership is everywhere.

I take great joy in my pre-flight routine of strolling through an airport terminal, perusing the bookstores. For some reason, these micro-bookstores brim with fascinating reads on leadership. Some of these manuals are academic, some are theoretical, and some are based on life experiences. As a captive consumer of the airport bookstore, I often load up on books that spark new leadership ideas. With each treatise I consume, I realize my life has been an incubator for leadership experiences. I will share with you intimate experiences of failures and wins that I hope bring clarity to your life.

Whether my audience is executives from the hotel industry, Sea-World, or priests from the Archdiocese of Washington, DC, I am passionate about sharing the lessons I have learned. Hopefully, everyone I edify takes something away from this process. For, I firmly believe that everyone is navigating the difficulties of life, and everyone is ***Navigating Leadership***.

I will share my experiences on the deck plates of combat ships, in the air over hostile waters, and in the corridors of power in Washington, DC so that you can benefit from my mistakes and successes.

As we travel together through the tapestry of leadership, remember that we cannot ascend to greatness alone. It takes a team of supporters aligned with your mission. I had a tribe of incredible Sailors and Marines. When I was off-duty, I was surrounded by a fantastic family who didn't volunteer to serve in the Navy. Still, I drafted my wife, children, and extended family into my adventure. Throughout this journey, my family met my experience with a loyal smile and unconditional love, never regretting their life in the Navy. They supported my career, made geographical moves, changed schools, and met every demand expected of a military family. And they, too, were required to lead and navigate the challenges that stood before them.

I have studied leadership in the classroom, instructed leadership in the field, and lived leadership in the real-world laboratory of the United States Military, commercial industry, nonprofit, and federal service. What I have digested, is not the end all be all, as I am still learning. Nevertheless, the lessons and life examples I am about to share have kept me in good stead. Adapt them to your life, and you will be off to a great start in navigating leadership and improving your everyday personal and professional lives.

Untether all your beliefs about success and becoming a student of your life. Practicing leadership adds quality to every aspect of your life and it improves the lives of those you lead. If you practice and improve the art of leadership, you will have the strategic tools to leave this earth better than you found it.

"The most challenging task in leadership is leading yourself." **True North by Bill George**

LEADERSHIP TAKEAWAYS

> ➤ Don't write anyone off until you find out what motivates them.
>
> ➤ Where you start does not determine where you will end up.
>
> ➤ We can be inspired by leaders everywhere, from a simple gesture to a book. We must be hungry to find leadership in all things.

ADMIRAL'S ADVICE

Never rule yourself out in your career endeavors. As I tell young emerging leaders, "Whatever you believe, you are correct. If you believe you will fail, you most likely will. If you believe you will succeed, you are correct because you will." Your thoughts program your actions. Never beat yourself up. Chaplain Don den Dulk always advised, "Don't should on yourself." Have no regrets. Move forward, move on, and lead.

CHAPTER ONE

THE END OF THE BEGINNING

"Leadership is the ability to inspire appropriate action beyond the expectable."
James Toner, Joint Force Quarterly

The historic halls of The Eisenhower Executive Building, EEOB, are the home to a lineage of leaders who offered support and guidance to the most powerful men and women during war, insurrection, and peace. As a result, every decision we make as presidential advisors provides a roadmap that bleeds into our nation's consciousness.

My training as a naval officer tempered the unpredictability of a National Security Advisor. The Navy offered me routine and stability. The structure of schedules, duties, and assignments quantified my role as an early leader. However, this rigid construct of schedule and duty never dulled my ability to amuse.

ENTER MY WIT AND WISDOM

Wit and wisdom have always been a constant in my life. I was hungry to learn, not just from books but from those who went before me. And my wit, as some may say, is an irreverent nod to be a little different from everyone else.

For example, during my tenure in the White House, I scurried past the security post at 0700 hours every morning (7:00 AM for my civilian friends) and was met by a very solemn protection officer who would say, "Good morning, Admiral Hall."

"Good morning, sir. I'm late again," I'd respond. "The President is expecting his foot massage at zero eight hundred."

I intentionally tried to counteract the pretentious air of snobbery many leaders brandish with a whimsical response. And, I admit, I was delighted when the often-serious officer would fight back a smile. I knew my wit had won them over. This is the way I connected with everyone.

My wit and wisdom became my secret weapon. This disarming approach to bridging a divide between the boss and the employee was so powerful that my adversaries would scratch their heads with bewilderment. My detractors didn't understand the arsenal of stupid dad jokes and mind-numbing one-liners that oozed from my DNA. Yet, my wit grabbed their attention. Looking back, I understand that my attempts at inciting laughter started the day I was born. I yearned for my parents' attention as I felt like the interloper in their rocky relationship.

NO MATTER OUR CIRCUMSTANCES, WE HAVE THE POWER TO LEAD

When you think of an influential leader, you automatically assume academic credentials follow. Yet, the blueprint for my leadership skills came to a screeching halt in the third grade when I was forced to repeat a year of academic hell. You read that correctly; I flunked third grade. Although I don't recall the name of my first third-grade teacher, I remember the woman who welcomed me to my second year as a third grader. Mrs. Paintner was an educator that saw beyond youthful antics and believed I was more than I appeared.

Up until this point, I was drowning in quicksand. My home life didn't support academic growth. At school, I was relegated to just a number. Unless you count wearing the crown of the most distracting student ever enrolled as an achievement.

YOUR PATH IS UNIQUE

Your path may be one of lost opportunity, hard lessons, and complex dynamics. Yet these are the qualities that define what true leadership is. A true leader understands that failure is the path to greatness. To navigate leadership, you must expose yourself to different experiences. Then, you must get out there and lead as though you have achieved the pinnacle of success.

One of my friends, Rear Admiral Sonny Masso, says, "One must seek to lead early and to lead often. Leadership is just like exercise; you need to do it on a routine basis to grow this muscle."

The good news is that you are already a leader if you're reading this book. You're either a designated leader or assumed to be a leader of your organization or team. You are looked up to by your team(s) and others. Whether leading people in the community, family, or organization, they want you to step up and manage your professional duties by teaching and making a difference. And so do I. Leaders will be rewarded beyond their wildest dreams. You have one chance at life. When you look back, will you have any regrets? Will you have been seen as a leader?

I prefer to be like my grandfather, a jovial chap who died asleep at the end of his life with a smile on his face. His final act was peaceful. He left this earth not screaming and yelling like the proverbial anxiety-ridden passengers in a car careening off the road. Instead, Grandfather lived his life without regret. He understood that we all have the power to navigate our own happiness.

THE PAST INDICATES THE FUTURE

A deeper look into the behaviors of past and present leaders presents a clear formula that is difficult to replicate. Historically, iconic leadership is premised on passion. This passion often transmutes to a life strangled with workaholic tendencies. Making a difference in the world must include being forward-thinking yet ready to adapt. There is a twist of luck that infiltrates the most famous of leaders. Being in the right place at the right time has catapulted many people into historical roles.

Sir Winston Churchill, the prime minister of the United Kingdom during the second world war, is an example of how passion, adaptability, and luck propelled an ordinary man into a unique position. I'm reminded of the quote attributed to Winston Churchill, who is believed to have said,

> *"The three hardest things to do in life are: first, climb a wall leaning toward you. Second, kiss someone leaning away from you. And third, talk about a subject that your audience knows more about than you."*

These simple words uttered by one of the most outstanding leaders of the 21st century was a seed planted in my leadership blueprint. This means that when I am met with obstacles, I embrace the challenge and never assume I know more than those around me.

16

CREATING A DYNASTY

I was not born into a dynastic family like the Roosevelts or Carnegies. I did not have forefathers who forged the way for my success. I had something more than nepotism; I had a spark. This intrinsic spark didn't stop me when people said *no*. This flicker of excellence helped me to deter naysayers. And it motivated me while enjoying no expectations from my lower-middle-class family.

Do you have that spark?
Do you know that you're meant for more?

Even playing curb ball in the streets of Buffalo, New York, gave me a tutorial on connecting with different walks of life. I wanted to be the best at everything I did. So, I created relationships with everyone from the drugstore attendant to the misfits on the street. Childhood gives you the gift of blissful ignorance, and I am delighted with the connections I made. I enjoyed the fruits of Buffalo's working-class chefs, the comfort of pizza, wings, bowling, and, most notably, cheering for the Buffalo Bills.

The backdrop of my story begins in Buffalo, a city that offers friendly, patriotic neighbors and families. The citizens of Buffalo are a robust and diverse community forged by lake effect snow and long winters. My people are known for their rally cries and for always supporting neighbors in need. This is the world I was born into and thrived in. My psyche was formed by good friends, great ethnic food, and, of course, The Bills.

I tested my chops as a leader by pursuing excellence as a teen. In high school, I was editor of my high school yearbook, co-captain of my varsity gymnastics team, and head of the cannon crew and spirit club. Following graduation from high school, I enlisted in the Navy with the mission and desire to attend the United States Naval Academy. Accordingly, I attended the Naval Academy Prep School after bootcamp. After one year of enlisted service, the Secretary of the Navy appointed me as a midshipman at the Academy. The four years there were the beginnings of an intense leadership laboratory.

The Naval Academy's mission is to develop Midshipmen by imbuing them with the highest ideals of duty, honor, and loyalty. They created leaders who had the potential to assume the highest command, citizenship, and government responsibilities. When graduation day arrives, a Midshipman is ready to lead on day one.

Following graduation from the Naval Academy, I entered naval aviation and spent 34 years as a Naval Aviator and a commissioned officer. Eight of those years were as a Flag Officer, or Admiral, in the United States Navy. Being a Flag Officer meant I was entitled to fly a flag that marked my position as the exercising commander. My banner depicted a blue field adorned with two silver stars.

Looking back, I realize that everything I experienced at every level became my personal laboratory for leadership. I studied the traits of notorious leaders and gleaned lessons from the good. I will share stories about these excellent and the not-so-good leaders. This way, you can determine what fits your personality and what to avoid in your leadership.

Leadership was, and still is, my calling. Leadership is about people, not things. As I learned in my military career, people are the Navy's true weapon system. As a lieutenant commander and a mid-grade officer, I worked for an admiral who told me,

> *"You do not command ships, you do not command helicopters, you do not command hangers or buildings, but you command and lead people."*

This is the one thing no one should ever forget. The purpose of leadership is to take care of people and lead them to a destination.

I once had an opportunity to speak with a venture capitalist who said that out in the business world, there are plenty of ideas and money. One thing that business lacks is the leadership to pull ideas and money together. The venture capital world is all about money and bringing ideas to life that will grow that money. Without leadership, that currency will not expand to its full potential.

WHAT IS THE EQUIVALENT CURRENCY IN YOUR FIELD?

Leadership affects the financial bottom line of a business, but what is the currency in other areas of life? In the military, specifically in the Navy, our currency is readiness, and you cannot achieve readiness without people who are professional and well-led Sailors and Marines.

Wherever we look, leadership is needed, including politics and our schools. Think of the situations we are facing as we regroup from the pandemic and remote learning. In marriage, leadership will teeter from one partner to another, but someone is always taking charge. This helps in raising strong, morally grounded families.

Friends chuckle when I describe my marriage to Darlene. I oversee the big decisions, and she is in charge of the small decisions. Before you gasp at my misogynistic approach to marriage–read on. In 45 years of marriage, I have yet to make one big decision. It does not always have to be one spouse or the other, but somebody always needs to lead the family forward. The momentum of moving forward is what anchors the art of tactical precision.

THE NOBLE CAUSE

Socioeconomic status does not preclude the emergence of leaders. If you're a wealthy mogul, this does not dictate the way you command a business. In fact, many leaders cannot be defined as righteous. Leadership lurks in the dark caverns of evil as well. Adolf Hitler is a prime example of leadership gone terribly wrong. He saw a need for struggling people, engendered trust with the public, and became history's most notorious mass murderer. Likewise, cartel leaders, drug warlords, and others with nefarious intent are individuals who lead in the wrong direction. The lesson in all of this is that if you aspire to greatness, you must have a noble cause.

A quick search of world news will reveal leaders on the world stage with anything but a noble cause. Their leadership may be effective, but it's heading in the wrong direction and leading to horrible outcomes. In 2014 I addressed the bullying style of a world leader, Vladimir Putin.

> *"International bodies are deliberating, talking heads are pontificating, politicians are politicizing, and only Vladimir Putin is taking action."*

The dissolution of the Soviet Union may have taken place in December 1991, but the Soviet style of action, negotiation, and bullying did not dissolve with the collapse of the USSR. It is alive and well with Putin, Russia's number one bully. Twenty-seven years of experience as a KGB agent, enforcer, and colonel have shaped the character of Putin. Phone calls from world leaders or sternly written letters from international treaty organizations will never change the bully.

My first experience with the Soviet bullying style occurred while on deployment as a Lieutenant Junior Grade in the US Navy. Our ship pulled into Mombasa, Kenya. During our customs brief by local authorities, we were cautioned about pirates and warned that even a mighty US warship in the Mombasa port is vulnerable to thieves. They pointed out that only the Soviets were safe from the pirate threat.

When I asked why the Soviets were safe, it was explained that during a visit by a Soviet warship, pirates attempted to board the ship in port, only to meet swift death at the hands of the Soviet sailors. The bodies of four pirates were dragged to the end of the pier and stacked up in a message to future pirates. This Soviet style of bullying was condemned by the host nation, and warnings were issued through official diplomatic channels. However, the only behavior that was changed was that of the pirates.

Twenty-five years later, I traveled to Lithuania, Ukraine, and Russia on official military business. Lithuania had recently joined NATO, Ukraine longed to join NATO, and the Russians despised NATO, calling it a 26-headed monster. Our group toured a KGB museum in Lithuania, which we quickly learned was a KGB prison, torture, and execution facility used to bully the Soviet satellite.

In the Ukraine, we learned the Russians were threatening to cut off oil and gas to the country if they continued their pursuit of NATO membership. In Moscow, we met with US diplomats and Russian members of Parliament. The US diplomats told us how Russian agents (using KGB style tactics) would enter their apartments while they were out. The agents would use their toilets without flushing and move furniture around. Nothing was ever taken other than the diplomat's feeling of safety.

Diplomatic protests would be made against the Russians. The Americans were bullied, and the Russian behavior was unchanged.

When meeting with members of the Russian Parliament, it was made clear that leaders longed to reestablish the Soviet Union and that they missed the Party. There we learned of their disdain for NATO. It was firmly stated they would no sooner allow the Ukraine to join NATO than America would allow Texas to secede from the United States. Ukraine contains Russia's strategic Black Sea Naval Base of Sevastopol. This invaluable port has been Russia's access to the Black Sea and the Mediterranean Sea since the 18th century. Therefore, the Russians will fight for access to this port.

Dr. Phil McGraw would say past behavior is a good predictor of future behaviors. Just as my 35 years in the US Navy have molded my behavior, character, and values, Putin's 27 years in the KGB have solidified his Soviet bullying behavior. He will never allow the Ukraine to join NATO or cede control of Sevastopol. He will continue his pursuit of returning Russia to a Soviet Union-style system like the one he grew up enforcing as a KGB colonel.

So as international bodies deliberate, analysts analyze, politicians pontificate, and world leaders affirm their disgust and disapproval, Vladimir Putin will act. Like the bully he has shown himself to be, he will have little regard for international concerns or threats of retribution.

Now is the time for leaders in the free world to put a stop to this bullying by putting teeth into their disgust and disapproval of his actions in Ukraine. The lesson to be learned is that not all leaders champion the greater good.

Reread this sentence and think about it. Understand that when you find yourself in the situation of addressing negative leadership, your responsibilities will increase. For example, you may need to combat foreign enemies, dismiss idle gossip, or even reprimand an energetic teenager. Whatever it is, you must be grounded in your purpose.

BOTTOM LINE: LEADERS MAKE THE TOUGH CALLS

This may be the most challenging action an elite leader may have to take.

Speaker Newt Gingrich (forgive the namedrop) told me during a discussion on leadership that:

> *"Politicians do whatever it takes to avoid pain, and leaders do what is right regardless of the difficulty of a decision. This piece of advice is one I will never forget."*

You must serve a greater cause; you must have the desire to serve others. Think of the leaders you know or have served with. Are they perfect? Absolutely not. Perfection isn't needed in leadership, but a noble cause is. Because you are reading this book, I assume that you aspire to lead your life with integrity. However, not everyone is like you or me. We must acknowledge that crack dealers, despots, and other evil ones also lead.

Today's newspaper or electronic news feeds show birth announcements in the social section and obituaries. In obituaries, you'll read about the difference an individual made to their family, community, and profession. Leadership accomplishments will probably be highlighted. However, birth announcements, list the proud parents and details of the birth. There will be no mention of the baby being a future leader. This is because leaders are not born, they are mentored and developed.

Therefore, my number one contention is that I can teach you to lead if you're of basic intelligence and motivation. If I can teach you to fly a helicopter, I certainly can teach you to lead. Now, just like a helicopter or an automobile, you cannot just be told how to fly one. You must get in the seat with a competent instructor.

You must get out there and do it. In other words, I can tell you that in a helicopter, the collective makes it go up and down, the cyclic makes it go fore and aft, and the rudders change your heading, but that's not enough. It's nice to know that information, but you have got to get in the aircraft and fly it to learn how to fly. It's the same with driving a car. Of course, you can learn to drive by watching a YouTube video, but until you get in, grab the wheel, and put your foot on the accelerator and brake, you won't learn how to drive.

Leadership is much the same way but is challenging to define. Here is what military author James Hugh Toner has to say on the subject:

> *"Leadership is the ability to inspire appropriate action beyond the expectable." So let me repeat, "Leadership is the ability to inspire appropriate action beyond the expectable."*

LEADERSHIP AND MANAGEMENT
ARE OFTEN CONFUSED

If you remember anything, remember this:

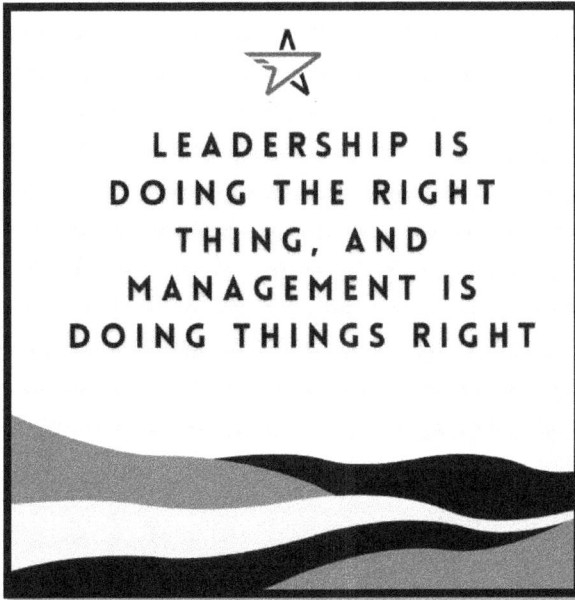

LEADERSHIP IS DOING THE RIGHT THING, AND MANAGEMENT IS DOING THINGS RIGHT

I challenge you to ask a preteen in your life to give you an example of someone doing the right thing, and then give you an example of doing things right. I proposed this same task to my son, Sam, when he was approximately ten years old. It's incredible to have conversations with your children while driving since you're both facing forward. There's no escape, and no eye contact is made. This way, you can have a very blunt and forward conversation.

"Dad, can I use an example from my health class today?" he asked.

Interesting. "Sure, Sam, go ahead. "

"If I was walking along a sidewalk and saw someone lying in the street with their head on the curb and bleeding profusely, doing the right thing would be to immediately run over to that person, stop the bleeding, ensure the breathing, and call for help."

Impressive. "So, give me an example of doing things right."

"If I was doing things right, I would go up to the individual. I'd first make sure I had a face mask on and rubber gloves, and I would ask them, 'Do you have any communicable diseases I should be concerned about? Is there anything that you are infected with that might harm me?' I would then stop the bleeding, ensure the breathing, and call for help," Sam said.

I was impressed with his answer and his grasp of the concept. Ask a young person these questions about the difference between doing the right thing and doing things right, and you'll be amazed by the answers.

> *Leadership is doing the right thing, and management is doing things right.*

If you trust the process, elite leadership is doing the right thing and doing it in the right way. Leadership contributes to the success of any organization on the margins and at no cost. Excellent leadership doesn't detract from the bottom line. In fact, leading with greatness is the vital ingredient required for any organization, whether it is a corporation, a school, a sports team, or even your family. Leadership inspires others to serve and to understand their potential in themselves.

Today's challenge in the American economy is retaining talented workers. People don't leave an organization; they leave their immediate supervisor. That's true in the military, and it's true in business.

Leadership makes a difference. It helps retain valuable people and brings profit to any organization's margins.

Leadership is a learned skill that must be practiced early and often. No matter if you're born in the heartland of America or Timbuktu, we all have access to improving ourselves. Leadership is a skill that requires you to navigate your environment, your people, and your life's situation. It is not always easy, but with a good chart and moral compass, you will find it easier than you thought. And boy is this rewarding.

LEADERSHIP TAKEAWAYS

> ➢ We need effective leaders (to do the right thing) and managers (to do things right). If you can do both well, your chances of success will dramatically increase.

> ➢ Failure is the cornerstone of doing things better.

> ➢ When doing the right thing, leadership "is noble" to differentiate between the Hitlers of the world and the Gandhis.

ADMIRAL'S ADVICE

Take complete ownership of your life and circumstances. Eliminate excuses of your upbringing or how perceived social status affects your current situation. Instead, find your noble cause, make the hard calls, and lead. Don't look back over your shoulder at your life. Instead, navigate by the stars, not the garbage in your wake.

CHAPTER TWO

THE SURPRISE PACT OF LEADERSHIP

"The greatest leader is not necessarily the one who does the greatest things. He is the one that gets the people to do the greatest things."
Ronald Reagan

A s I noted, I was not born to be a leader. In fact, the odds were stacked against me. I was born in a lower-class family to unhappy parents and was a poor student. So, when I recognized that I had an opportunity to commit to bettering myself, I chose the PACT.

WHAT IS A PACT?

A **PACT** is an agreement, alliance, and, notably, a covenant that shall not be broken or forgotten. An elite leader has a PACT with excellence. When choosing this PACT, you act with Passion, are Accountable in all things, remain Committed throughout, and know the Traits and Traditions of influential leaders.

LEADERSHIP PACT
PASSION
ACCOUNTABILITY
COMMITMENT
TRAITS & TRADITIONS

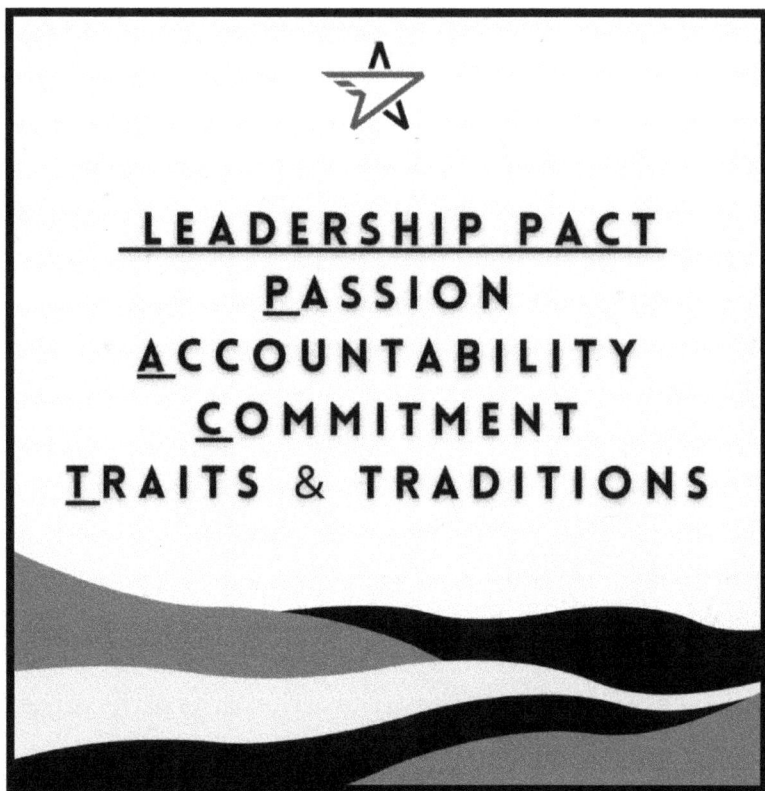

THE HISTORY OF THE PACT

My career began with the grandeurs of an officer and the goofiness of an attention-starved kid from New York. Four years at the Naval Academy laid the foundation of my leadership training and education that chipped away at my rough edges. It instilled a keen focus and new dreams I could never have articulated back in Buffalo.

As new graduates, 800 of my classmates and I were eager to hit the fleet and serve our commitment to the Navy and the Marine Corps. I signed my name on the dotted line and gave my life to serve my country. Having been selected for naval aviation, my service commitment clock wouldn't begin until I was designated as a naval aviator.

The service commitment never concerned me because I was focused on earning my Wings of Gold and being the best naval aviator. Off and running like a rock star, I joined my roommate, callsign Sweens, on our Fall Flight School start date. The challenge was navigating Flight School and our newfound freedoms post Academy life.

Flight School started in Pensacola, Florida, and the competition was fierce. Walking in on day one, I could feel and taste the competitive fervor. Getting deeply involved in the training and getting the best flight grades was imperative. The hunger for being the best was rewarded with the best squadron seats in the fleet.

Sweens and I were the only two academy graduates in our Ground School class. The other students came from Reserve Officer Training Corps or ROTC at civilian colleges. Or they graduated from the officer candidate school, OCS, a three-month course. If they make it through the 90 days of training, we refer to them as our 90-Day wonders. Our four years of intense Academy life made Sweens and I the natural leaders of our cohort. Being young and bold, our leadership capabilities were met with many distractions. We had the world at our fingertips, promising careers, and youthful vigor. Yet, we chose to invest in a mission more significant than ourselves.

RISKY LEADERSHIP

Case in point: At the start of ground school Student Naval Aviators (SNAs) were required to take a battery of engineering exams to determine our strengths and weaknesses in our technical education. Flying multimillion-dollar naval aircraft takes a great deal of knowledge of systems, both aerodynamic and weapons employment. These exams determine if you need additional technical training. For example, a Liberal Arts Degree in Comparative Poetry from a state college would put you at a disadvantage in aviation.

Much of our class wanted to form a study group to prepare for the several-hour exams we would face the next day. Confident in our recently bestowed technical degrees, we convinced everyone that the exams would be a "piece of cake" and "pastry." So, we headed to Barrels Bar for nickel beer night, followed by a trip to the beach for a nightcap at the Foul Deck Bar and Grill. Sweens and I led, and they followed.

The next day we took our exams through the fog of beer goggles. After the exam, our class walked out, heads hanging low and giving Sweens and me side-eye glances that telegraphed their anger and disappointment.

"Piece of cake," Sweens shouted.

"Pastry," I added.

We were leading but not in the best way. I hadn't made a **PACT** with leadership yet. Later that week, the results were published, and only two class members passed all the exams and weren't required to take additional training. *Sweens and Hall.* We passed with flying colors, and our reputation was set – follow these two at your own risk.

There were three primary training squadrons in Pensacola and one primary squadron in Corpus Christi, Texas, at the Naval Air Station. An announcement was made at our roll calls that two students needed to go to Corpus Christi for continued training to balance the flight training load.

Sweeney and I looked at each other and shrugged.

"What the heck?" he said.

"Let's go," I added.

Maybe we were looking for an adventure in Texas. Or perhaps, we instinctively knew we had worn out the welcome with our current classmates. Either way, having no idea how our decision would change our lives, we loaded up my old Volkswagen camper and were off to Corpus.

BECOMING A POSITIVE LEADER

At the time, there was a backlog of students needing to be trained. The new primary flight trainer aircraft, the *T-34C*, was behind delivery schedule. The intermediate trainer, the *T-28 Bravo*, *The Ensign Eater*, a mammoth radial engine bird, was filling the aircraft gap. Keeping these aircraft maintained and in-flight condition added to the backlog. Combined, these events resulted in an aviation training pool and a backlog of student Naval Aviators.

Students met only once a week in uniform for roll calls and the dissemination of information, and then we flew. In the meantime, we were expected to execute 30 training modules on meteorology to be ready for cockpit flight training. After four years of intense academic rigor, we chose the beach over the classroom, thinking we had plenty of time for everything. We focused on barbecues, happy hour parties, and the sun. We were living a worry-free good life.

Sweens was engaged to his high school girlfriend, and I thought I'd never be married because I'd be too busy with a life of flying and adventure. Then one night, at a student pilot Beer Fest and cookout, Sweens met a young co-ed and the beach beauty he would eventually marry.

A few weeks later, I met Darlene, a petite hometown gal with a strong family and community structure. Of course, it did not hurt that she was also a knockout. It was love at first sight for me, but she had excellent reasons for not marrying a Navy guy. Determined to be with her, I deftly balanced my time between flying and dating while I navigated my life, burning the candle at both ends.

Despite her original misgivings, we were married in a traditional Catholic wedding six months after we met. Three days after we married, we packed into my faded red VW van and headed to Pensacola, where I finished my flight training.

Life changed, and my new bride would proudly pin on my new Navy Wings of Gold. Graduating number one in my flight class gave me the first choice in picking a fleet squadron. Following six months of additional flight training in my fleet aircraft, we were headed to Barbers Point, Hawaii, on permanent change of station orders. We traveled with our small dog, who kept Darlene company during her pregnancy.

It was time for me to grow to the next level of maturity. I was a husband, an expectant father, and now an operational fleet pilot ready to deploy at a moment's notice.

Darlene was a brilliant new mother, competent and in charge. She was thrown into a life she had little knowledge of, and she adapted quickly. Her Texas upbringing and friendly personality made her a favorite in the squadron. Her strength at home enabled me to focus on learning the tactics, techniques, and procedures of being an anti-submarine warfare pilot. She was the rock of our family and, in true Navy form, gave birth to our son while I was at sea.

When asked why she was so optimistic when faced with the challenges of Navy life, she always replied, "Life is always a choice, and I choose to be happy for myself and my family."

BECOMING A NAVAL INSTRUCTOR

In the three years of that tour, I spent over 18 months flying at sea. As a result, I became a tactical and technical specialist. I enjoyed leading the Navy helicopter detachment's small teams of sailors. We wrapped up our tour in Hawaii, where Darlene excelled in supporting the squadron spouses and young sailors, and I excelled in the air. Since the squadron's commanding officer thought I would be valuable as a flight instructor, he sent me to the Fleet Replacement Squadron in San Diego, California, where I would instruct new pilots, preparing them to enter the fleet just as I'd done three years earlier.

I relished the role of instructor. I was now a seasoned Lieutenant preparing newly minted naval aviators and air crewmen for their tour in the deployable fleet. Racking up hours improved my flying skills. And the work on the ground and in the ready room impacted my development as an officer, aviator, and standardization and check ride pilot.

I spent as much time talking and giving oral exams in the classroom as I did educating pilots in the air. As instructors, we engaged in chair flying missions and quizzing our pupils on emergency procedures and the unusual scenarios they might encounter. It was this banter and the back and forth that sharpened our knowledge. The oral exchange is a tried-and-true method across all communities. This involves techniques such as challenge and replies in checklists, Q&As on procedures and tactics, and debates over far-fetched scenarios.

Finally, we enjoyed the humorous sea stories and retelling of old and odd adventures. Our motto was, "Never let the truth get in the way of a good story." And this is deck plate learning and leadership.

LISTEN AND LEARN

I was growing my brand and improving my skills and, in the process, becoming a little overconfident and impressed with myself.

As a team, Darlene and I were always up for squadron fun. One evening we were invited to a dinner party with the more senior officers in the squadron. Skipper, our commanding officer, was one of the guests.

As the dinner progressed, the conversation turned from polite, casual talk to a more serious tone. I was the junior man at the table and became the focus of a discourse on leadership. I fell back into my plebe year role of fielding tough questions during meals from upperclassmen. *Be confident, don't equivocate. Don't wander; get to the point.* It hit me that I was in an oral exam on leadership.

It was three against one, and our silent wives slid their gazes back and forth as if watching a tennis match.

I held my own, providing points on leadership. I also took away insight from the seniors, who had more experience than me. This became an inflection point that made me think, act, and respond.

When the evening ended, I was mentally exhausted and exhilarated by the give and take. I needed to espouse these words of wisdom from my superiors. The most important being that you can have all the technical skills in the world, but you need to be a leader to be effective. If you can label and categorize your leadership vocabulary and thoughts, you can improve and execute them. If you improve your leadership, you'll improve your life. After all, leadership begins with you.

My PACT with leadership was born over a Navy dinner party. When writing your own PACT with leadership, you will gain the tools to navigate your organization's challenges, vocations, and personal life.

LEADERSHIP TAKEAWAYS

> ➢ Determine your PACT in your personal and professional life.
>
> ➢ Lead early and lead often. Leadership is like a muscle - exercise it.
>
> ➢ Fortune favors the bold. Get out there and do it!

ADMIRAL'S ADVICE

Action in life is like a Naval maneuver; success is assured if you start early and in the right frame of mind. Find your passion early in your personal and professional life, and you will be on fire with success. When landing on a ship with your aircraft, you'll have a smooth landing if you approach the ship on altitude, course, and airspeed. If you are off on any of these parameters, you'll struggle to land. Figure out the parameters of your job and lock them in early and often.

CHAPTER THREE

PASSION AND PURPOSE

*Nothing great in the world has ever
been accomplished without passion.*
Georg Wilhelm Fiedrich Hegel

assion is described as a strong and barely controllable emotion
that drips into your veins with the cadence of a hurricane. Passion
can drive a steadfast leader to unimaginable heights, or it can knock
you to your knees, asking God, "Where did I go wrong"? The moment
I met Darlene Allred, I knew my passion for success would be moot
without this force of nature by my side.

In 1977, I was a happy-go-lucky naval officer fervently attracted to
good times, classic rock, and captive audiences. In retrospect, I craved
the attention I hadn't received as a young boy.

I attracted audiences with my love for magic tricks. Stopping a
passerby on the street and performing magic wasn't unusual. My go-
to routine would involve pulling a quarter out of a stranger's ear, then
waiting for their elated response. The look of delight in a stranger's
eyes confirmed I was worthy of love and attention.

I was at the height of my bachelorhood during the 70s. This time
period of generational changes fostered the growth of social angst, 8-

track tapes, hitchhikers, bell-bottoms, and disco. It was an excellent time to be alive.

The unbridled feeling of youthful vigor along with what I considered my striking looks, would be the ticket to memorable times. I anticipated my well-earned liberty (the time off from military duties granted upon missions accomplished). Taking liberty from the Navy meant I could blow off steam and share my burgeoning wit with whomever accepted it. This particular break called me and my partner-in-crime, Sweens, to the mecca of spring-break-paradise, Corpus Christi, Texas. More specifically, Trini's Jolly Beggar. The popular joint was a pirate-themed watering hole smoldered in a haze of cigarette smoke and stinging with the pungency of stale beer.

Something tangible filled the air that night. My feeling was confirmed when a beautiful blonde-haired, blue-eyed girl strolled in with her cousins. Dressed in my best, I knew luck was on my side. Freedom from the intense military routine left me with an exhilarating buzz.

Clad in a blue denim jumpsuit, Darlene sat in the corner, sipping a white wine spritzer. Radiating southern-belle charm, her energy pulled me in.

I excused myself from my present company and edged my way to the woman who would one day become my wife.

"Hello, I'm Garry," I said, overflowing with confidence.

The petite Darlene was at least a foot shorter than me. The freckles on her shoulders stole my breath. Her piercing blue eyes gazed into mine and her lashes fluttered with the charm of Scarlett O'Hara. When the corners of her lips crooked upwards, I knew she would be mine.

"Hello, Garry," she responded, her tone indifferent and not at all matching her friendly body language.

"Would you like to dance"? I asked

"No," she said. "You have a reputation for being unkind. You know what I'm talking about....the jean jacket?"

"Sorry," I murmured as I recalled lending one of her friends my jean jacket then demanding it back.

At the time, the rumor mill had churned with colorful stories of my arrogance. In my brief time at Trini's Jolly Beggar, I had become infamous for my magic tricks and cocky demeanor.

While gazing into those stunning eyes, I exhaled. I needed to take drastic actions to fix my reputation because this beauty wasn't getting away.

Aiming for humbleness, I said, "I'm not the man that you think I am." I unbuttoned my jacket and ruffled my hair. "I'm a nice guy and would like your phone number."

After looking me over, Darlene frowned. "I am only eighteen."

Not knowing that she had lied and was twenty-one, I backpedaled and rethought my attempt to connect with the gorgeous woman.

From minute one, Darlene pulled me into her ecosystem, playing hard to get with ardent force. She didn't know it at the time, but I love a challenge. In fact, *challenge* is my middle name.

This is where the superpower of my wit paid off. Combining fearlessness with compassion, and adding a heaping dash of humor, I dug my heels in and made Darlene my next mission.

I exited the bar, taking on *Mission Darlene* with a new swagger. Afterall, I held her phone number scribbled onto a beer-soaked napkin. I have never looked back. This was my destiny.

DARLENE

Darlene was raised in the family structure I craved. Her parents, Felita and Sam, led with love. And love became the currency that fed their three children.

Felita provided strong familial ties to their Mexican heritage. Legend has it that within the annals of a dusty history book, Darlene's great-grandfather rode horseback with Pancho Villa. Their rich Mexican history is bound to mothers cooking homemade flour tortillas to accompany the aromatic picadillo.

Sam was born of hearty Oklahoma stock, bringing stability, structure, and an abundance of love. They built a modest life in Corpus Christi that entwined mutual adoration, familial values, and long nights on the porch sipping iced tea.

Darlene's parents were a staple in their Corpus Christi neighborhood. Each night they perched on the front porch, chatting with neighbors and watching bustling groups of children play in their front yard. It was an idyllic upbringing fostering Darlene's steadfast ability to love.

When we met in the spring of '77, Darlene was spoken for. She was knee-deep in a relationship with "Mr. Wonderful," a dapper heir to an impressive used car dealership. On paper, he was ideal husband material, checking all the boxes. He wore a three-piece suit and walked with indignation. He was ambitious, steady, and had a crown of hair that was perfectly coiffed. Moreover, "Mr. Wonderful" could speak Spanish, which impressed Darlene's family. However, he lacked the wit of a scrappy young naval officer who promised adventure sprinkled with moments of intense passion. It was this passion that fueled my relentless pursuit of Darlene.

I instinctively knew that if I was going to become the man I was meant to be, I needed a compass to navigate my way. Darlene became this compass. In fact, she's been my compass for over forty-five years leading me to become a loving husband, father, son, and friend. She built a home filled with affection and joy that I lacked for my first twenty-five years.

We all need an anchor, no matter our trajectory. We'll lose ourselves if we don't have a solid community of people who love us. Luckily Darlene agreed to be my anchor, and we married six months after we locked eyes at Trini's Jolly Beggar.

The passion in my personal life was now fulfilled. We were young and in love. We had no idea what was ahead in our lives, but we knew it would be a fun adventure with laughter and the sound of future children. Darlene was barely twenty-one, and I was twenty-five with a chest puffed with self-importance. Looking back, I ask myself "What were we thinking at such a young age?"

A few years ago, we hosted an Irish Catholic Priest in our DC home for a week.

When asked about our life and how we met, he interrupted us. "Twenty-one and twenty-five? I never would have married you two."

By the end of his visit, he had changed his tune declaring, "Now that I've spent a week with you, I take back what I said. I would have married you."

I fully believe that our marriage came to pass because of the intervention of a higher power; we were meant for each other. Forty-five years later, older and wiser, we have a home filled with laughter and passion.

My passion in my professional life was growing. With my ally Darlene neatly tucked into military life, I yearned to make my mark as a Naval officer. Coming from a childhood family with little love and affection, I was passionate about making a difference with Darlene and our life.

First, I wanted to honor her and make her proud of my professional career. Second, I desired to be the best pilot in the squadron to prove she married the right man. My profession became my second passion. I did whatever it took to win Darlene, and now I was all in to do what it would take to be the best leader and aviator in my squadron.

FROM HUSBAND TO HERO

As passionate as I am for Darlene, I am equally passionate about the people I serve. Passion for your spouse, family, and loved ones is easy to understand. It's a foundational emotion that leads to outstanding achievements. Indeed, I have a spouse that elicits my passion–for her, my family, friends, Sailors, and Marines. Leadership is a calling, and passion is required to be an elite leader and navigate your life. To be an outstandingly elite leader, you must have passion.

Alabama college football Coach Nick Saban says we have five choices to make in life:

- *You can choose to be bad at what you do.*
- *You can choose to be average at what you do.*
- *You can choose to be good at what you do, which is what should be expected for your God-given talents.*
- *You can choose to be excellent, or you can choose to be elite.*
- *To be elite at any endeavor takes desire and passion.*

You can choose to be bad at what you do.

You can choose to be average at what you do.

You can choose to be good at what you do,
which is what should be expected for your God-given talents.

You can choose to be excellent, or you can choose to be elite.

To be elite at any endeavor takes desire and passion.

-Nick Saban, University of Alabama Football Coach

SIGNING UP FOR THE NAVY

When I started my college education and serving our country, I wanted what today's young men and women want: adventure, education, opportunity, and the association with like-minded individuals desiring to serve and lead. I wanted to be a hero.

The Navy offered a path to these desires. Besides, I was ready to bolt and wave goodbye to Buffalo and shoveling snow. The Naval Academy promised a life of adventure, professional acumen, and the ability to differentiate me from the lot I left behind.

In Naval Aviation, there were basically two distinct types of pilots. As a young lieutenant, I was a Squadron Schedules Officer. This was my second squadron, and we were responsible for training pilots to fly in the fleet squadrons. Production of qualified fleet-ready pilots was paramount to our mission. In addition, we had a cadre of instructor pilots who recently came from the fleet to accomplish the task.

Taking a newly minted Navy pilot fresh from flight school and qualifying them in a complex mission aircraft was no simple task. It takes motivated instructors with real experience. This is where my two different pilot tasks came into play. There were the instructors who said unconvincingly that they wanted to fly training missions and there were the pilots with every fiber in their body that wanted to fly the training missions. These distinct responsibilities require different characteristics. Saying is one thing and doing is another. Talk is cheap; actions speak louder than words. Those who said, again unconvincingly, that they wanted to fly were called seagulls because you had to throw a rock at them to get them off the ground. The pilots that absolutely without exception wanted to fly took the tough flights, the problem students, the early morning flights, and the late-night flights.

The seagulls were the first to hit the snivel log, the logbook of excuses, such as why they couldn't fly before mid-morning or after dark. I told my instructors that scheduling them was like dealing with cookies from a big jar. Some cookies were whole and delicious, and some were crumbling and stale. So, if they'd take a broken cookie today, I'd ensure they got a warm cookie the next time. Pilots that took any mission built their qualifications and became elite pilots.

Pilots who were the first to snivel or whine about their given mission usually made the minimum flight hour requirements and made average pilots. Elite leaders have a passion for leading early and often. They make tough calls. They lead, not just say they lead.

If you don't have the passion and desire, you need to think long and hard about how to develop it. Because if you are a reluctant leader, it's a long march to the head of your organization.

LEADERSHIP TAKEAWAYS

> ➤ Passion and desire go hand in hand. You must want to lead.

> ➤ Nothing worthwhile is accomplished without passion.

> ➤ If you don't have the desire to lead, it's a long march to the head of your team.

ADMIRAL'S ADVICE

Take inventory of your passions and desires. Do they take you in the right direction? Would you be willing to discuss them with your grandmother or priest? If your answer to both questions is yes, you're on track to be an elite leader. Embody desire and enthusiasm, and you'll lead with great success.

CHAPTER FOUR

ACCOUNTABILITY
– CAPTAIN OF YOUR SHIP

Accountability can only rest with one individual and cannot be delegated. That person owns the consequences of the team's performance. Therefore, accountable leaders inherently earn the respect of their team.

> *"A relationship, I think, is like a shark, you know? It has to constantly move forward, or it dies. And I think what we got on our hands is a dead shark."*
> **Woody Allen, in the movie Annie Hall**

S harks must constantly move forward immersing their gills with oxygen-rich water. As a leader, you, too, must progress forward. One way to do that is through continued education. Leaders are life-long learners.

As an instructor pilot on shore duty, I enrolled in an executive MBA program conducted on base by Southern Illinois University at Edwardsville. One of my academic advisors at the Naval Academy once told me that my Marine Engineering degree, accompanied by an MBA, would be a powerful combination.

Darlene was saddled with more responsibilities than me. Our three young children tugged at the hems of her apron, demanding food, homework help, and bedtime stories. Despite how busy she was, Darlene selflessly allowed me this opportunity for growth. Without a supportive partner, your chance at obtaining meteoric excellence is almost impossible.

While Darlene led on the home front, I was determined to make something more of myself. This may seem like my ego drove the boat, but I knew my kids, Garry, Courtney, and Sam, would watch their parents' sacrifices and emulate them in their own lives.

The grueling schedule of work, study, and classes, topped off with a few hours of sleep, depleted me of the energy I needed to be the *Leave it to Beaver* dad I had craved as a young boy. However, when I cracked my books open, I knew my education would benefit my family in the long run. After all, we must make sacrifices for the greater good.

One evening, early in my studies, I realized that these business courses, coupled with my extensive background in engineering, would be the key to unlocking my true potential. By the dimly hung light of our kitchen in Rancho San Diego, my lightbulb went off. The pages of my textbooks gave me the answers. Management, production, and promotion would play a big part in my future successes.

At the time, this rigorous academic schedule, demands at work, and being a family man seemed overwhelming. The more I improved the more I demanded of myself. It was as if I was on a never-ending treadmill.

I lapped up everything I could in my Executive Management course. This class, taught by a salty retired Navy Captain with searing eyes and pursed lips, impacted me in a manner reminiscent of good

old Mrs. Paintner, my third-grade teacher. And luckily my highly struc-
tured days still allowed for a few *aha* moments.

A particular Saturday session sits neatly in the back of my mind,
making its appearance whenever I'm in stressful situations. The salty
retired Navy Captain pounded his fist on the lectern, exclaiming, "The
number one trait of executives under stress is blaming others."

Blame: (definition) Assign responsibility
for a fault or wrong.

Aha. All too often, leaders at any level under stress blame others
for their circumstances. This is due to a lack of accountability. The angry, blaming archetype doesn't have the anchor of accountability.

It is the path of accountability that separates ordinary leaders from
great leaders. The great leader accepts accountability without equiv-
ocation. The concept of accountability is a difficult one. It's often
equated with responsibility which is also important in leading others.

Any PACT made is only words on a piece of paper if there is no
accountability. This means you can talk the talk, but you must walk
the walk. To understand this, the best examples are found in the sea
services. The captain of a ship is in a position of ultimate accountability. As long as sailors have gone to sea in ships, the captain is held
accountable for that ship's safe navigation, morale and combat readiness.

While I was in high school, I worked for a seasoned retired Petty
Officer, Joe Smeller. When he saw my interest in the Navy, he shared
that when at sea, the chain of command goes Captain and then God.
In the Navy, if something catastrophic were to happen, even if the
captain is asleep in the middle of the night, he or she is held accountable. They take this role with the utmost regard and understanding.

For example, suppose the ship has sailed into shallow waters. In that case, the navigator may be responsible, but the captain is held accountable. This is the ultimate "buck stops here" in leadership.

Responsibility may be delegated, but accountability always remains at the top with the leader. Responsibility and accountability go hand in hand, and you must understand both to be an elite leader. When the going gets tough, the tough get going. They never blame others. While we were plebes (freshmen), we were given five basic responses. As eighteen-year-olds, we may not have realized at the time we were being taught accountability. Our primary responses were:

Yes, Sir/Ma'am
No, Sir/Ma'am
No excuse, Sir/Ma'am
I'll Find Out
Aye, Aye, Sir/Ma'am

"Aye Aye" is a traditional response in the sea services. It means *I understand and will obey or carry out the instructions given to me.* At no time were we to use the phrase, "I don't know." There was also an unwritten law, *We will never bilge a classmate,* a nautical saying that means never throw your classmate under the bus.

As eighteen-year-olds, we may not have realized we were being taught responsibility and accountability. However, this is exactly what was happening. We were accountable for our actions, education, and the understanding of our situation.

The act of accountability is not for the faint of heart. Accountability takes courage and self-awareness. Courage is never the absence of fear but rather having the skills and preparation to press on despite present fear. Those you lead in your organization can easily smell any lack of accountability at the top. Once you have lost their confidence in your ability to accept responsibility, you have lost your team's best performance.

President Lincoln, in a letter to General U.S. Grant, said, "I neither ask nor desire to know anything of your plans. Take responsibility and act and call on me for assistance." Accountability and responsibility are often used interchangeably when they are two sides of the same coin but never interchangeable.

The elite leader recognizes their responsibilities to their profession and to their personal life. An emerging leader leads a small group in an organization and must be a technical expert in their field. They're responsible for knowing their role in carrying out the organization's mission.

For example, a young Naval Aviation Petty Officer is highly trained and must be a technical expert in maintaining a multimillion-dollar aircraft. That's their responsibility. The Junior Naval Aviator must be a master of their aircraft to fly and fight this lethal machine.

At every level, regardless of position, one has professional responsibilities to ensure effective efficiency. This area of responsibility is easily understood and measurable. However, personal responsibility is more difficult to account for. You cannot lead if you don't take care of yourself and take responsibility for your behavior. You are the one responsible for your health, mental well-being, work-life balance, and personal conduct.

You are always on stage as a leader, and everyone's watching. Those you lead will observe how you behave in both good and bad times. If you aren't a person who behaves responsibly, the weight of leadership will be difficult to bear. When leading, your personal life is exposed and your private world cannot be separated from your professional world, no matter how good you are.

I served on an amphibious assault ship with a crew of 1000 and a contingent of 2000 Marines. Our second in command previously commanded a different amphibious ship. He was an expert in amphibious operations, the deployment of Marines, and the safe navigation of Naval vessels.

On paper, he was professionally outstanding and seemed the perfect man to serve as second in command. Unfortunately, he was also lacking personal accountability..

In off-duty hours, he participated in the underbelly of our society, engaging in activities that are not morally acceptable in most cultures. Living a double life eventually crossed into his professional life, and the officers and crew of that ship quickly realized his lack of personal responsibility. Morale declined, orders were followed begrudgingly, and the entire performance of the vessel deteriorated. In the end, the officer could not keep up the façade. An investigation was conducted, and he was removed from Naval duty.

The ship slowly recovered and conducted a successful six-month deployment, but not without the challenges associated with recovering from the lack of responsible leadership by the second in command.

Suppose you aspire to lead and are required to lead. In that case, you must do so with a balanced and moral life, both personally and professionally. When navigating life and leadership, you must step up, take responsibility and lead. Your team is not only counting on you. They are also watching you.

LEADERSHIP TAKEAWAYS

> ➤ Ultimate accountability remains with the leader. Effective leaders also create a culture where each individual understands that they are accountable for their actions.

> ➤ Only responsibility may be delegated, Accountability is never delegated.

> ➤ Elite leaders lead a moral and balanced life.

> ➤ Know and develop your responsibilities.

ADMIRAL'S ADVICE

Take accountability seriously in both your personal and professional life. You'll sleep better, your life will run smoothly, and you'll be happy. Don't blame others or shirk your responsibilities. Instead, be accountable for all your actions. When you make a mistake, own up to it quickly and succinctly. Do not provide long stories about the path to your mistake as it only shows weakness.

CHAPTER FIVE

COMMITMENT TO THE MISSION

"People do not follow uncommitted leaders. Commitment can be displayed in a full range of matters to include the work hours you choose to maintain, how you work to improve your abilities, or what you do for your fellow workers at personal sacrifice."
Stephen Gregg, Chairman and CEO of Ethix Corp.

S tephen Gregg's quote on commitment resonates with me. When I want to convey a commitment to an audience, I point out that commitment is like an All-American breakfast. This always elicits tilted heads and strange looks as the audience wonders where I'm going with this comparison

LEADERSHIP FOUND ON A DINER'S GRAND SLAM

Your average All-American diner breakfast is two eggs any style, hash brown potatoes, toast, and bacon. Of course, we've all had a restaurant's Grand Slam breakfast or something similar to this high-caloric treat at least once in our lives. You know that breakfast that's loaded with heaps of fried meat, eggs any way you want, and shoestring potatoes soaked in a puddle of glorious grease. This breakfast, and its eclectic deliciousness, is the perfect analogy to leadership. Let me explain.

It takes a diverse team to create this magnificent meal. Combine a chicken, a farmer, a baker, and a pig and you have a perfectly plated breakfast. The chicken is the matriarch providing the eggs. Our Idaho farmer planted the seeds that turned into crispy potatoes. A local baker mixed the dough and baked the bread, that becomes the toast slathered in butter. And that poor pig dedicated his life to making the best bacon a guy could eat. The pig's selfless commitment to the greater good wins Porky the greatest accolades.

BE THE PIG

To lead well and decisively, you must be committed; you must be the pig in this breakfast. A commitment is dedicating yourself to your organization, its people, and its mission. You're all in, focused, and motivated if you're committed. This pig-like commitment and your moral compass give you unwavering direction in leading. You never second-guess your decision to be the perfect accompaniment to the Grand Slam. You know your mission in life, and you are unstoppable. Commitment to your position is what keeps you going when challenges arise. Commitment is the consistent application of leadership.

OFF TO JAPAN

Overseas life is not easy for a service family. Learning basic language skills to survive was only the tip of the iceberg. Darlene was tasked with finding living quarters and schools where our three kids would thrive. Homesickness was a common thread in our adventure in Japan. Everything was new and strange, and situational depression was a commonplace. Wiping the tears from my small children's eyes as they begged for their favorite foods and ached for their friends proved to be an unexpected challenge. I often questioned my choice to uproot my family again, as they supported my rising star.

The personal challenges my Sailors and Officers encountered in moving to Japan, combined with the professional challenges we had to overcome in obtaining the squadron's combat readiness, put a strain on every leadership level. From division chiefs and petty officers to division officers, department heads to my executive officer and me, the stress was palpable. It is commitment that sustains you through the challenges, the highs and the lows and leads your organization forward.

My previous boss, Vice Admiral Eytchison's voice rang in my head as he reminded me, "When in command, you do not command aircraft, hangars, or ships. Never forget that you lead people." In any mission, never forget you are committed to the job and the people you lead.

Being the pig in that breakfast means you are the ultimate example of commitment. Commitment keeps you steady on your course and leading with purpose, regardless of the highs or the lows, the rewarding days, and the not-so-rewarding days. And, if you are a vegan, be that inextricable piece of seasoned tofu that knows why and who you serve.

THE PIG IN JAPAN

The day finally came, and I was about to establish and command a Navy Seahawk helicopter squadron at the Naval Air Facility in Atsugi, Japan. Only one in five Navy helicopter commanders was selected for command so I was over the moon. An overseas command in the Pacific required leadership and a great deal of diplomacy.

The day was filled with excitement. Our vintage World War II hangar was decked out in full-dress Navy style. Naval nautical flags, state and country flags, and the Japanese national ensign adorned the ceiling and walls. Our first aircraft was stationed near the dais and was all spit and polish. Dignitaries were assembling in the green room. There were twelve Japanese Admirals and Captains and a handful of U.S. Pacific command Admirals and captains. This was a big deal.

A beautiful Darlene arrived clad in a peach colored linen suit, nude heels, and our youngest son on her hip. Beaming with pride, her blue eyes welled with tears as she watched the military pageantry. The Navy band played prelude music and the Sailors, in their crisp white uniforms, proceeded in formation. Their intense patriotism exuded collective awe from onlookers.

After a while, Darlene approached me. "I'm missing part of my brood. Where are Garry and Courtney?"

"They are in the green room entertaining twelve Japanese Admirals," I called over the ambient beat of the drum corps.

"No, really, where are they?" she asked.

I crooked my finger, and she followed me to the holding room, filled with dignitaries. A twelve-year-old Garry spoke Japanese as he poured cups of coffee for the guests. Courtney, ever the angelic smiling eight-year-old hostess, passed out freshly baked pastries while nodding politely at the comments made in a foreign language.

My entire family realized the importance of this day and had risen to the occasion. They were part of the Grand Slam. Each one did their part to make the meal delicious. Unbeknownst to them then, they were mini ambassadors positively impacting our foreign allies.

This day was an emotional high. The pageantry, the accolades, the music, the food, and my family coming together for the common good is a memory seared into my mind. Everything went off without a hitch, and Helicopter Anti-Submarine Squadron Light Fifty-One, The Warlords, was established the first week of a new fiscal year. It is effortless to be committed to the mission on days like this one.

Unfortunately, the moment was fleeting. The formal uniforms, neatly pressed and without a stain, adorned with medals and swords, would be replaced with fatigues and flight suits. Once the music faded and the dignitaries left, it was time to get to business. It was time to move forward, move on and meet new challenges. We embraced our new chapter.

LEADERSHIP TAKEAWAYS

> ➤ Your first big decision is to be committed to your team.
>
> ➤ Commitment takes courage and keeps you focused through leadership's emotional highs and lows.
>
> ➤ Be the pig. All in or nothing.

ADMIRAL'S ADVICE

In all activities, you choose, be committed. Lack of commitment is easily exposed in a profession, a position, or a relationship. Commitment is a force multiplier and a powerful trait in all group settings, whether professional or personal. And finally, when interacting with the troops, your spouse, or your children, put the smartphone away.

CHAPTER SIX

THE TRADITIONS, TRAITS, AND BEHAVIORS I LEARNED WHILE NAVIGATING MY PROFESSION AND LIFE

Humility has not come easily to me, and this was perhaps my biggest challenge in crafting my PACT with leadership. As you can tell from the previous chapters, I indulged in arrogance. A life built on the illusion of grandeur is a life that is easily destroyed. An inflated ego and sense of self-importance, and a need for attention and admiration lead to distrust and disgust. Living an ego-centered life repels those closest to you, makes you critical and competitive, and leads to frustration.

Understanding the traits and traditions that align with your leadership style is imperative to accelerating your future. For example, if you value integrity and compassion, being an egomaniac diverts your path to success. My blueprint for success did not include being pretentious. However, for a short time, no one would know differently. The lesson of humility has been the saving grace in my life. I would not have learned this lesson alone. I was graced with good people who cared enough to tell me enough is enough.

Let's take a quick look back at my rise to leadership. I was not born into a lavish lifestyle or lucky enough to have parents who backed me. My father told me I would fail out of the Naval Academy, and I proved him wrong.

On top of that, I fell passionately in love with a Texan beauty who met me with stoic rejection–*I changed her mind.*

Despite many rejections, I was intent on proving I would be successful. I believed I had the mental capacity to be anything I wanted to be, and I was entranced with the possibilities of my future. Of course, I had become madly in love with *Garry, Garry, Garry.* (More on this later)

This endorphin rush from being the man of the hour and hero consumed my thoughts and puffed up my chest. Quite clearly, I thought I was awesome. So, when I finished a successful squadron tour as an officer in charge of a deployed helicopter detachment, and department head, I took a promotion I was certain I could handle. Afterall, I was born to hobnob with the rich and famous. I was ready to be the phenomenal Garry I imagined. I knew I was on my way to the top and I would kick ass and take names.

I was recommended to the Bureau of Naval Personnel to serve as an aide to a senior admiral. *Boy, oh Boy, was I on fire.* I was a tactical expert with proven skills. *I'm a natural.* While deployed, my team exceeded all previous accomplishments of an early helicopter detachment. *Because we were awesome.* We set goals, hit the targets, and achieved them while flying more days and nights than expected. We escorted tankers through dangerous waters, chronicled our tactics in professional journals, and earned the Secretary of the Navy's Helicopter Ship safety award for the entire Navy. *Check.*

That year I was selected by the Naval Helicopter Association as the Pilot of the Year for sustained performance. I was the obvious choice; I was the top gun and ready to rule the world. Obviously, we were good, and I was pretty full of myself. I knew I could handle any job in the Navy. I was on fire so who wouldn't want me on their team?

Clearly, being an admiral's aide would be a cinch.

Boy was I wrong.

AN AIDE TO THE ADMIRALS

I was told upfront, and with absolute veracity, that I would be sent to work for the most demanding admiral in the Navy. As an aide, there were two paths–you excel, or you're fired. After a one-on-one interview with the admiral in which I answered his questions on leadership, I was hired. *Probably because I was fantastic.* Although, legend has it that I was hired because I declared loyalty was my number one priority.

The tour lasted thirty-three months and ten days, but who's counting? I served in the Midwest and supported two admirals who turned out to be as different as night and day. The first admiral was proud of his reputation and a tough son of a gun. He was brilliant in nuclear tactics and techniques but lacked softer skills. The second admiral was the consummate gentleman and statesman. He was just as competent technically but had the personal skill set beloved by all, especially his spouse and children.

I learned a great deal from both men. From Admiral Eytchison, I learned the balance between work and life, family, faith, and service. He was wicked smart, as proven when he was selected to head the investigation into the nuclear accident at Three Mile Island. He was equally brilliant about life, travel, and leisure. While traveling, we often shared a fine meal and would top the evening off with a cigar and the best scotch whiskey.

Once or twice a month, Admiral Eytchison and I traveled from the Midwest to Washington, DC, for meetings in the Pentagon. In DC, we used the visiting Flag Officer offices and stayed in temporary quarters in the Washington Navy Yard. The Navy's Chief of Naval Operations (CNO) lived at this base.

As we were carrying our luggage to our quarters on one summer trip, we strolled along the wrought iron fence that circled the CNO's garden. The admiral halted and set down his bags. He reached over the fence and plucked a long rose bush stem to his nose.

He inhaled and smiled. "That's something, Commander, have a whiff."

Oh, my goodness. The man in charge of developing our nuclear war plans stopped to smell the roses. This waypoint in my career was never lost on me.

I learned two lessons that day in the CNO's garden. The first is that your actions speak louder than your words and often show true character. This powerful man showed the balance his humility gave him. Second, as a leader, you are constantly being watched.

What did I learn from the first admiral? You can be a son of a gun and demand the best out of your team, yet you will not be building your successors. True leaders create other leaders, not followers.

I also learned that the devil is in the details and to check and re-check my plans and actions. I developed tremendous skills serving this admiral. I also learned when there is no balance in your life, life can be hell.

Working for a tough boss is like navigating through a storm. Sometimes stormy weather makes for a better mariner, and working with an SOB boss sometimes can make you a better leader.

I worked with tough-as-nails Marines, Navy SEALS, and seasoned Mariners and found that the elite leaders had humility about them. Humility doesn't equate to weakness but to introspection into one's strengths and capabilities and how to use them to best lead a team.

As I have said, humility didn't come easy to me. I was pretty arrogant in my youth. My priorities were *Garry, Garry, Garry*. And, as Darlene often pointed out, I was all about "Garry and Garry." I was beyond self-centered. After all, I was the Navy's Pilot of the Year and Aide to a powerful Admiral. Everyone wanted to fly with me, deploy with me, and pull liberty with me.

Unfortunately, I was forcing myself to be successful. I was not humble, balanced, or doing well in my family life. I was being presented with the lessons I needed but was oblivious to their importance.

DARLENE PROVES TO BE THE ELITE LEADER

The tour came to its conclusion. I had served two three-star admirals, learned tremendous lessons beyond flying, was selected to the rank of commander early, and was chosen to command a new squadron in Japan. I was ready for the next level of leadership.

Still, with all this success, something was missing, but I could not pinpoint the issue. Now I understand, I was out of balance, my priorities were all about me, and clearly, the lesson of humility had not yet taken hold.

GARRY, GARRY, GARRY

I was blind to my faults, but Darlene wasn't. She saw through my false bravado and felt an uncomfortable twinge. Nevertheless, she plodded on, caring for our kids, leading military wives through their challenges, and came home to a man whose self-importance trumped everything. After all, I was proving myself as an infallible leader in the Navy. I navigated the aircraft efficiently and sat at the table with the most influential people in the world. Damn, I was incredible!

However, one day I pulled into the driveway after a long stretch and was met with sad blue eyes. Something wasn't right.

Darlene has an intuitive intelligence beyond compare. She saw the big picture of our lives together and craved a smidgen of the familial stability she'd grown up with. Her macho husband Garry was absent, full of himself, and wearing the badge of pride on his chest. Truth is, I was stretching our marriage thin, and she was fighting to keep it together. Her love was fierce and unconditional, yet, she had the fortitude to know when she was on a sinking ship. It's essential to have an anchor like Darlene in your life, someone who has your best interests at heart and will tell you the truth.

Darlene was bogged down in child-rearing duties and felt she was essentially a single mother. She tirelessly gave of herself to better the lives of our little family, with the optimistic slant that her husband was changing the world. However, her husband was cocky and self-serving and minimized her leadership. She told me that I would find my way into shoal waters and be forced aground in my professional and personal life if I didn't get it together. This was the wake-up call I needed.

MY COME TO JESUS MOMENT

We left my Midwest tour with a new baby. Darlene wasn't ready to move overseas with three youngsters. She knew my life was unbalanced and I was full of myself. So, she took on the mantle of leadership and challenged me with tough questions about my priorities. She saw this cocky helicopter pilot rising in the ranks and focused on success while ignoring his most significant responsibilities at home.

I remember this day with clarity. We were living in San Diego, California. I was preparing to start a squadron in Japan and undergo training in the new aircraft. Courtney, our middle child and daughter, was preparing for her first communion, and Darlene was making all preparations for an overseas family move. Although she effortlessly maneuvered the logistics, she struggled with my dismissive attention to our family. Tensions were high as I worked through Darlene's questions.

"Are you going to be a leader, or are you going to be one of the boys?"

"Are you leading to inspire or just for the thrill of being important?"

"Do you understand the importance of your family as a leader, or are we simply adornments, like the medals on your chest?"

Her words cut into my soul and made me rethink everything I knew about leadership.

In 1990, we were a couple in crisis. It feels like yesterday when we attended Courtney's First Communion rehearsal at a beautiful Spanish mission-style church. This adobe structure with a circular clay roof had asymmetrical facades, large square pillars, and a bell tower that flanked the main building.

The large sanctuary was the iconic backdrop for our little girl's communion. However, the tension between Darlene and me was palpable. We sat midway back at the first break in the pews. Darlene's arms crossed her chest, and I leaned away from her. Our body language didn't go unnoticed.

A woman approached us from behind, threw her arms around us, and pulled us together. She then stated with a calm positivity that the

Blessed Mother, Mary, appeared in the back of the church and instructed her to tell us that everything in our marriage would work out.

I get goosebumps to this day, retelling that event. We had no idea who this woman was, and we have not seen her since. However, she gave me the wake-up call of all wake-up calls. It wasn't enough to be told our marriage would work out. I had to take the helm of my life and get my priorities balanced.

I sought counseling, worked hard at introspection, and realized my intense love for Darlene. Before heading to command in Japan, a new set of priorities replaced *Garry, Garry, and Garry* with God, family, and the Navy, in that order. I never looked back. Because of my wife's plea for her family, I altered my course and led in a more robust direction.

My obligations to the Navy didn't diminish. Still, I had a new perspective and better balance when making decisions for my professional and personal life. I was navigating life and leadership with a set of charts. Previously I was fighting and forcing myself to be successful, and my number one priority was me. Now that my priorities were aligned and balanced, my career took off like a rocket. Our new squadron won all the professional awards, morale was at an all-time high, and I was ranked as the number one commander in the Pacific.

The lesson on the traits of humility and balance is like any lesson. You are never one and done. You must practice humility and balance over and over again. And never forget, when you lead, everyone is watching. As the old saying goes, the higher the flagpole you climb, the more your rear end shows. So don't be an ass.

HUMILITY TAKES STRENGTH OF CHARACTER AND THE COURAGE TO LET SUBORDINATES SEE THE REAL YOU

Leading starts with the individual. You must be able to lead yourself before leading others. Elite leaders practice ongoing introspection every day of their lives. Thinking you have your job, marriage, and life squared away isn't being introspective. I keep returning to the quote from the movie Annie Hall: "A shark must constantly be moving forward, or it dies.

Achieving a marriage close to five decades long takes introspection and constant improvement. Leadership requires the same work. In a marriage, humility makes it easier to be loved. When you lead an organization, humility makes it easier to follow. In both cases, your actions speak louder than words.

I met with Father William Byrne a few years after my retirement from the Navy. My corporate life didn't require the same time commitment as active-duty Naval service, so I wanted to continue to serve in some capacity. The church felt like the right fit.

After a noon Mass, I spoke with the junior priest in our parish, the Parochial Vicar. I said I was ready to volunteer in any capacity, from pulling weeds in the Mary Garden to doing handyman work in the church building. He was a Navy reservist and knew of my background, so he smiled and indicated the Pastor may have better things for me to do.

A week later, I met with the pastor, who greeted me with, "You must be the Navy guy with time on your hands."

I smiled and nodded a *yes*.

Father Byrne explained that he had been praying for a leader to come forward for the past three months, so someone must've answered his prayers. This was another example of divine intervention in my life.

Father Byrne said he wanted to start an adult ministry bible study in the homes of Capital Hill parishioners. He'd provide the material, and I would supply the structure. He had the vision, and I had the desire, so I committed.

Father Byrne, now Bishop Byrne, said I was the answer to his prayers. However, I believe he was the answer to mine. We worked well together. I led the ministry for four years. Darlene and I were then asked to lead a module preparing young couples for marriage. And on several occasions, Father Byrne asked me to speak on faith-based leadership and life-work balance. Together we advised couples on the power of marriage.

Byrne's leadership was needed in another parish, and when he left, he said, "I'm now your spiritual advisor. Come see me once a month." Our connection was sealed. Byrne became an integral part of my team.

Two months passed after he left Capitol Hill, and we bumped into each other at an event in DC. He asked with a furrowed brow why I hadn't come to see him yet.

"Aye Aye, sir. I will schedule a visit asap," I said with a chuckle.

At our first meeting, he asked about my family, how I was doing, and where my spiritual and prayer life stood. Unfortunate, I bobbled my response on the spiritual life, so he said, "Let's develop a plan."

I confidently said, "I'm a world-famous helicopter pilot, callsign Viper, and I know God has an amazing plan for me." I was trying to bring humor into a delicate conversation.

He smiled. "I agree with you, and I can prove it."

"Can you?" I asked with a grin.

"Yes, I can. You're still alive."

LEARNING POINT

If you're alive, you still make an impact. You're still navigating life and leadership. You always have a purpose if you're on this side of Heaven.

Together we came up with a spiritual growth plan. Before I left that first session, Byrne said he had something for me and printed off a document. He handed me a folded sheet of paper and smiled. "You'll be needing this. Let's work on your humility as well as your prayer life."

Father Byrne had given me the Litany of Humility, a prayer I carry with me always. It's powerful, and it hits me in so many ways. Any leader of any faith can benefit from this Litany.

Humility of A Hero

The Litany of Humility

Given to me by Bishop Byrne

O Jesus! meek and humble of heart, Hear me.
From the desire of being esteemed,
Deliver me, Jesus. (repeat after each line)
From the desire of being loved, From the desire of being extolled,
From the desire of being honored, From the desire of being praised,
From the desire of being preferred to others,
From the desire of being consulted, From the desire of being approved,
From the fear of being humiliated,
From the fear of being despised,
From the fear of suffering rebukes,
From the fear of being calumniated,
From the fear of being forgotten, From the fear of being ridiculed,
From the fear of being wronged,
From the fear of being suspected,

That others may be loved more than I,
Jesus, grant me the grace to desire it. (repeat after each line)
That others may be esteemed more than I,
That, in the opinion of the world,
others may increase, and I may decrease,
That others may be chosen, and I set aside,
That others may be praised and I unnoticed,
That others may be preferred to me in everything,
That others may become holier than I, provided that I may become as
holy as I should.

I include The Litany of Humility, a Catholic prayer, first given to me by Bishop Byrne because he has been my mentor for years, and every line of this prose reflects different phases of my life.

1. I desired to be loved as a child.
2. I desired to be extolled in my teenage years.
3. My desire to be extolled and praised was fed during my tenure in the Navy.
4. My two years in the White House satisfied my desire to be consulted.
5. And writing this book confirms my desire to leave a legacy.

Yet humility granted me the opportunity to better navigate life's challenges. Moreover, humility gave me the greatest gifts: Peace, Grace, and Understanding.

LEADERSHIP TAKEAWAYS

➤ It is crucial to have an anchor like Darlene in your life, someone who has your best interests at heart and will tell you the truth.

➤ You are only as successful as the team that surrounds you.

➤ Humility is not a weakness; it's a strength.

➤ Our ego exists primarily in the external world and acts like a shield to protect our need for status, self-worth, and contribution.

➤ Believe in something greater than yourself and have faith.

ADMIRAL'S ADVICE

Take the time to observe leaders in your organization. Determine who you feel is an ethical leader that operates with good character, and then seek them out as a mentor. Never be afraid to ask someone to mentor you. As my father-in-law would say, "what are they going to do? Hit you?" Being asked to mentor a subordinate is a high compliment. And remember Ethel's advice; All things in moderation.

CHAPTER SEVEN

THE CALLING TO BE SOMETHING MORE; HEALING FROM LEADERSHIP LESSONS

I was not the husband I wanted to be. I was on the path of recreating the generational trauma of my parents. We are all handcuffed, in some sense, to the teaching of our families of origin. I knew I wanted to be more. I knew that because Darlene believed in me and gave me the support I needed, I owed it to her to shed the shadows of my past. I needed to do the work. I needed to heal. I needed to step up and lead my family as I was meant to. Leadership starts in the home.

After six months of intense spiritual studies, daily Mass, and attending retreats, I met Father Byrne. "I know what God wants me to do," I said.

He loosened his collar, leaned back in his oversized chair, and signaled his beloved dog to join him. "I'm looking forward to what you've learned."

"God wants me to pick my clothes off the floor and put them in the hamper. Next, he wants me to empty the dishwasher and trash without being asked. He wants me to be tender to Darlene, and finally, he wants me to be prepared."

The pastor grinned. "I think you're finally getting it."

No more *Garry, Garry, Garry.*

Leaders are humble, considerate, and prepared. However, I didn't realize how this process prepared me for the unknown events that lay ahead. I still had more leadership to navigate.

The elite leader has humility at the heart of their character. This humility allows you to share with your team and subordinates who you are, and to ask for their support.

TWO YOUNG SAILORS GIVE ME THE ULTIMATE BOOST

At anchor off the coast of Yemen, with the smell of coffee and salt in the air, the *Tarawa* rocked, dancing with the rough seas around her. It took significant wave action at forty-two thousand tons to make her roll, telling the mariner in me that, despite sunny skies, the ocean was making its power known.

The crew went about the business of the day. Acting as the command ship in the rescue and recovery of the *USS Cole*, each department performed its specific role flawlessly. However, our mission off the port of Aden was coming to a close.

The *Cole's* crew had performed with great honor and courage following the terrorist bomb that left seventeen of their Shipmates dead and a gaping hole in the side of her hull. They fought tirelessly to keep their ship afloat and tend to the dead and wounded. Everything in their power was brought to bear, and the ship was saved.

My engineers assisted by ensuring the life-saving shoring held up and relieving *Cole's* exhausted crew. *Tarawa* and two destroyers helped The *Cole* with meals, showers, and a resting place. Through *Tarawa's* robust communications suite, *Cole's* Sailors could call home to loved ones longing to hear their Sailor's voice. I could not be prouder of the American Sailor. Their resilience, spirit, and professionalism were eye-watering.

The Cole had been floated out to the Russian drydock ship earlier in the week and was now high and dry and ready to be taken back to her home port in the US. We would be transporting its crew to Oman to be airlifted back to the States ahead of their wounded ship.

On this day, our Commodore, Colonel of Marines, and I would travel to the destroyers to thank them for their support of *Cole*. Then the three of us would visit the Russian ship carrying *Cole* home. The Russians were the only country with ships capable of lifting a destroyer out of the water and transporting its cargo at great distances. They have two ships, the *Black Marlin* and the *Blue Marlin* and the *Blue Marlin* was offered up for this mission.

The current sea state and *Blue Marlin's* lack of a helicopter pad meant the three of us would travel by Navy seven-meter rigid hull inflatable boat or RHIB. The RHIB was crewed by three seasoned Sailors, a coxswain, an engineer, and a boatswain's mate. We entrusted them to handle this seven-meter craft that can achieve 40-plus knots with skill and professionalism.

With a trill of the bosun's pipe over the ship's announcing system came the command, "Man the port side cargo hatch for boat ops!"

This was my call to join the Commodore and Colonel on the port side of *Tarawa* and meet our ride for the day's mission. We would thank the other ships, visit the *Cole* and ensure she was secure on the *Blue Marlin*.

The uniform for this mission was a fresh set of heavy khakis with a long sleeve shirt and ballcap. Not the ideal apparel for the one-hundred-degree heat and 85% humidity the Gulf of Aden was offering up.

Adding to that uniform, we would be donning that less-than-fashionable, crotch-grabbing floatation device called the kapok. It was designed to keep an unconscious sailor afloat, head up out of the water, and to be easily visible with its bright orange color. The kapok was not designed for comfort or maneuverability. But per boat operations safety procedures, we would wear them even if alternate floatation devices were available.

Every Naval evolution has its customs and traditions. Regarding boat operations, the senior officer always boards last and departs first. That put the boarding order with me first, the colonel second, and the commodore last.

From the cargo hatch down to the RHIB, the distance varied 20-30 feet as the seas dropped and rose at its own rhythm and whim. The drop was straight down and accomplished by a Jacobs ladder. A wood and rope ladder painted bright orange was a standard means of boarding and debarking ships worldwide. Using the Jacobs ladder requires grip strength, balance, and steady nerves. An older SEAL officer reminded me that fear gives you greater grip strength, which is an excellent point to remember.

The RHIB was close aboard, and my deck crew secured the bow with a painter delivered from *Tarawa*. Why it's called a painter escapes me, but the RHIB was secured to *Tarawa* and rising and dropping with the seas. I've been here before. Grab the side of the hatch while a Sailor steadies you, spin around and face the hull, and head down without hesitation. All while knowing that the boatswain's mate on the RHIB will grab you wherever he can and pull you into the boat.

He tugged on me and, *Touchdown!* I was in. I took my position on the port side of the RHIB, grasping the holding rail on the center console.

Next at bat was the Colonel. He was right out of central casting, athletic, and with a chiseled jaw. He was one hundred ninety-five pounds of twisted steel, muscle, and cat-like reflexes. After a quick brief, he came down the ladder, and the RHIB rose to meet him. Once he was aboard, he took a position opposite me, grabbing the rail on the center console.

Two-hundred sixty pounds of Commodore came into place. The Colonel and I exchanged grimaces. "If he falls on us, he could kill us both," said the Colonel.

To our safety, he landed soundly in the forward end of the RHIB. Then, looking only forward, he took a Washington crossing the Delaware pose, entirely in character with the pompous assurance we had come to know.

The cockswain, a deeply tanned Petty Officer with an anchor tattooed on his right forearm and a ship's wheel tattooed in the crook of his left thumb and forefinger, was our seasoned and confident driver. The boatswain mate released the bow painter, and we were free from the mothership *Tarawa*.

The cockswain looked me in the eye and said, "Captain, smooth or fast!?"

"Fast," I replied, knowing that was the answer the crew wanted. They loved showing off the RHIB's capabilities, and the Commodore loved a rough sea.

In a shot, we were off, and the Commodore was thrown back. Luckily, he landed firmly on the gear locker, forward of the center console, arms and legs flailing. The Colonel and I smiled as we gripped the handrails tightly, flexed our knees, and held on as the RHIB heaved into the waves. We circled to the left, counterclockwise, to cut behind *Tarawa* in route to our destination.

Later the air boss, who was observing from the tower, said he watched in awe as the RHIB jumped off wavetops. He reported we were entirely out of the water and airborne at one moment.

Our visits to the two U.S. destroyers on the scene were uneventful. Their smaller displacement meant they rose and fell with the seas, unlike the massive *Tarawa* that barely moved. With the destroyers and the RHIB in sync, it was easy to dismount the boat and jump on the destroyer's Jacob's ladder, and with two steps, we were on board.

We thanked each ship that came to *Cole's* aid and bid them safe sailing as their mission off Aden was now accomplished. The destroyer's captain offered a traditional VIP Navy lunch of club sandwiches and iced tea, followed by freshly baked chocolate chip cookies.

Our chariot, the ship's Ridged Hull Inflatable Boat (RHIB), and her crew were summoned alongside, and we were quickly off to visit the *Blue Marlin* with her cargo of the wounded *Cole*.

It was as if we were on a wild carnival ride. We just didn't know when or if the ride would end safely.

The *Blue Marlin* grew more prominent as we got closer to her anchorage. She was massive, and from a distance, *Cole* looked like a child's toy sitting on the *Blue Marlin's* deck. But, like *Tarawa*, the vast hull of the *Blue Marlin* was not tossed about by the rolling seas. Instead, the ocean rose and fell along her sides with a thirty-plus foot of change. Thoughts of boarding her in these seas produced incredible grip strength as I held tightly to the center console.

Our radio crackled with the voice of a Russian crew member instructing us to approach their starboard side, where we would see their Jacobs Ladder to board.

I scanned Marlin's starboard hull, looking for that crisp international orange ladder that all ships carry. *Nothing!*

When the coxswain flicked his head toward Marlin and shouted to me, "There she is!" I didn't see it, so I scanned again.

A dark and dingy mess of wood and line hanging on Marlin's side came into view.

What the heck?

Every step was slanted in a different direction. One vertical line holding the steps was longer than the other, and both terminated in what looked like old dish rags.

"This should be fun," I whispered even though my stomach churned.

Under normal sea states, this ladder could be handled with a good grip and speed. However, on that day, the ocean decided to add an additional degree of difficulty by rising and falling approximately thirty feet at an irregular rhythm. One moment, the sea was 20 feet from the bottom of the ladder, and the next, the powerful water was swallowing four of the ladder's eight steps.

The Commodore assumed his Washington crossing the Delaware pose again as we approached *Marlin's* hull. One foot positioned on the inflatable gunwale, the other planted on the RHIB's deck. The boatswain's mate's steadying hand sat on his orange Kapok life preserver. As the RHIB's bow kissed the hull, the seas lifted us halfway up the ladder, and the Commodore quickly did a grab and go. In a heartbeat, he was up and on the Blue Marlin. In that same heartbeat, we fell with the seas 30 feet below the ladder, but passenger one was safe on *Marlin*.

Next up was the Colonel. He assumed the same position the Commodore had only moments before. The seas lifted us up, but the Colonel hesitated, waiting for the perfect exit. He grabbed the vertical lines of the ladder, but the seas fell before he could catch one of those crappy rotten steps with his foot. He was hanging from the ladder, the sea 25 feet below him. The seas came back up and swallowed him up to his thighs.

As the scene seemed to move in slow motion, the powerful water released him.

I thought for sure the next swell would take him away. Still hanging by his grip, the Colonel scrunched his body into a fetal position, the next wave barely missing him. Then with a guttural Marine Corps, "*Arghh!*" he pulled himself up. He remained in a compressed body position but managed to catch a step with his boot. With impressive agility he ascended the ladder to the deck of the *Marlin*.

The mouths of those watching the feat of human strength hung agape.

I swallowed hard. *My turn.*

The crew of the RHIB waited for my move. Aiming for calm, I projected my voice to the coxswain. "Fall off for a moment."

We circled about 200 hundred meters away from the *Marlin*. I motioned to the three-man crew around me calling them to the center console.

"We just witnessed a feat of human strength only a superhero can summon. But unfortunately, I can't do that. I need you to get me on that ladder without hesitation."

Three heads nodded in unison, one Sailor saying, "We got you covered, Captain."

That statement coming from one of my burly Sailors still induces goosebumps. I loved my Sailors, and their actions proved they reciprocated those feelings.

I assumed the position, left foot planted on the deck, right foot on the gunwale. This time the boatswain's mate was on my left and the engineer on my right. Their respective outside hands steadied me by holding onto my kapok. It didn't raise my consciousness that their inside hands were clenched in fists because I was focused on the Jacobs Ladder as the coxswain approached *Marlin's* hull skillfully.

His timing was perfect as he judged the rising seas. He came to kiss the hull precisely as the seas peaked, and he yelled, "Now."

Two strong fists contacted my buttocks in that split second and launched me up the ladder.

I was on deck before I could say, "Wow!"

When asked what I had said to the boat crew, I responded, "I wanted to point out I wasn't superman like the colonel."

After the men chuckled, we spent the next hour touring the *Blue Marlin* and surveying *Cole* while studying how she was secured to the deck.

The hour passed quickly, and the seas rose and fell 25 to 30 feet. I was the first to reboard the RHIB. With a thumbs up from the coxswain, I swung onto the excuse for a Jacob's ladder and started my descent. The RHIB rose to meet me, and the two Sailors who launched me were now there to grab me and haul me safely into the boat.

All three grinned at me. One firmly slapped my back, and the other shouted over the engine noise, "We got you covered, Captain."

I gave them two thumbs up. We were shipmates, and there is no greater love than one shipmate for another.

With the Colonel and Commodore onboard and in position, the coxswain looked at me, and before he could ask anything, I said, "Fast." With a team of smiling crewmen and the roar of an engine, we headed back to the *Eagle of the Sea* with her 1,000 Sailors and 2,000 Marines. Our mission for the day was complete.

TRUST IS A MUST FOR EFFECTIVE LEADERSHIP

I highly recommend *The Speed of Trust* by Steven Covey Jr. Trust can be difficult to earn and easily lost. Covey's book should be read by anyone who desires effective and efficient organization. You cannot serve as a commander in any role without knowing, liking, and trusting. A leader skilled at building and supporting good relationships is one of the main drivers of trust. Without trust, you have nothing.

You earn trust by executing the attributes discussed earlier in this book. For example, being a technical expert shows you're responsible, humble, and committed. In addition, your ethical style inspires trust and confidence. When you see your team members emulating these same attributes, you also learn to trust them. This works in both directions.

In the Indian Ocean, specifically off the coast of Somalia, piracy became a thriving industry that had tremendous pay off for the thieves. Cash was the new pirate's bounty. Earlier, they would steal radios and other equipment from ships they boarded without opposition. These pirates then forced shipping companies to hire security teams and to sail miles off their regular routes for safety. This added tremendous costs that were passed on to consumers.

At this time, the United States Central Command was determined to deter piracy and get shipping back to normal. To do this, the pirates' primary means of operating needed to be destroyed. That meant sinking or destroying their fleet of skiffs, the boats they used to chase down ships and board them.

Orders came down to the Fifth Fleet and from Fifth Fleet to my staff of Expeditionary Strike Group. We were to take care of the fleet of pirate skiffs by sinking or destroying those boats.

The commander of the Fifth Fleet had trust in me. So, at our usual daily intelligence brief, he gave me the simple order. "Take care of this, Admiral Hall. "

"Aye aye, sir," I said.

Back at headquarters, I gathered my staff. I'd worked with these individuals for a year and a half and trusted them implicitly. I laid out the situation from the morning's briefings. I told them we were tasked with eliminating the pirates' skiffs and that this fell under our rules of engagement of terrorism at sea.

The team embraced the task and set the plan in motion. They coordinated the rules of engagement and communicated the plan to the US Navy ships near Somalia, who we trusted as outstanding mariners and professionals. That day, as the sun went down, several of the pirate's skiffs were destroyed, and their burning hulls glowed on the horizon as they sank into the sea.

The entire chain of command in that operation trusted the organization, and a mission was ordered, planned, and executed in one business day.

When I communicated to our fleet commander the results of our mission, he asked that I pass a Bravo Zulu to the entire team. A simple Bravo Zulu is a Navy signal meaning *job well done*. Often the only recognition a Sailor wants is a, "Well done, Team."

The speed of trust in action.

Months later, after smooth sailing for commercial fleets, the pirates rebuilt a small flotilla of skiffs, and actions picked up again. I'd moved on after a standard tour length, and my year-long deployment was over. Unfortunately, my replacement was not a trusting individual by nature. When the call came to sink the skiffs, the lack of trust from the staff slowed the process. It was several days before the operation was completed.

Build trust and identify trust, and your
organization will operate efficiently
and at no additional costs.

EMPOWER TO GROW SUBORDINATES

To lead is to imply you're going somewhere and taking others with you. You're navigating to a destination. Sometimes different routes accomplish the same outcomes and at the exact costs. An elite leader doesn't operate on the principle of "my way or the highway."

Give your team clear guidance and then let them execute in their way within your general boundaries. You'll be surprised at their ingenuity and growth in confidence.

Getting a large ship like the *USS Tarawa* underway and out to sea takes two to three hours of working through the underway checklist. Engineers activate equipment, sailors secure equipment, and deck hands prepare the lines. Meanwhile, bridge personnel work to get everything ready for the team to sail out of the harbor.

One captain didn't trust his ship handlers, nor did he empower them. These ship handlers were junior officers eager to qualify to drive the ship.

Knowing that the captain would give them every order during the underway out of the harbor, they had little desire to prepare for the evolution. Everyone from the sailors at the helm to the petty officers in the engineering room felt they were parrots repeating the captain's orders.

Sure enough, as we got underway, the captain gave the junior officers every turn heading and speed change. As a result, it took them a long time to learn how to handle the ship throughout a deployment.

They never felt empowered and thus never felt motivated, and empowerment is the key to accelerated learning and growth.

Finally, my day came, and I was the captain of my own ship. I was determined to empower a crew of 1,000 men and women and 2,000 Marines.

The first time I got the ship underway from San Diego harbor, I entered the ship's bridge at the beginning of the two-hour checklist. I called the junior officers to the navigation table and asked the navigators to join as well.

As they leaned in and listened, I explained that I'd be sitting in the captain's chair observing as we got underway. I'd be there to back them up and provide help, but I wouldn't tell them how to sail the ship out of the harbor. Instead, I told them to sail us out, plan accordingly, and not hit the Coronado Bridge that spans San Diego Bay.

I left the bridge, reentering about 45 minutes later, and what did I see?

The junior officers crowded around the chart table, sweating profusely as they planned the course, speed, turning points, and landmarks that would get *Tarawa* underway. Watching from my chair, I couldn't help but smile as they worked as a team to safely get us into that deep ocean off the coast of southern California.

As the watch ended and the next team entered the bridge to take over, the young officers and Sailors high-fived each other, smiling with a sense of achievement. I was beyond proud of what they had accomplished and how they'd grown in one four-hour bridge evolution.

Trust to lead with speedand empower
others and watch them grow.

LEADERSHIP TAKEAWAYS

> ➤ Strong and effective leaders are individuals imbued with the trait of humility.
>
> ➤ Operate with courage, develop your trustworthiness trait, and learn to identify the trust in others.
>
> ➤ Empowering your team for growth is a force-multiplying behavior.

ADMIRAL'S ADVICE

There is a saying that a rising tide lifts all boats. Develop the traits and behaviors of a leader who empowers others and recognizes and rewards subordinates. Be humble about their success, as no one achieves success without the team they lead. Do this, and you will be the tide that lifts all boats.

CHAPTER EIGHT

NAVIGATING ROUGH WATERS

"If there is a flaw in your personal character, your leadership will quickly reveal it."
Anonymous Navy Admiral

My interest in navigation began in junior high. I was good at math and science and was fascinated by numbers and the potential scientific theories promised. An astute English teacher saw this, and when I balked at an assignment that required reading a biography, she suggested I dip into *Carry on, Mr. Bowditch*. I was hooked on the hero's story and imagined Nathaniel Bowditch's indentured life as an accountant's assistant. Nathaniel, like young me, was talented when it came to math and numbers. After years of hard work, his indentured service paid off, so he followed his call to sea.

During this timeframe, the maritime world was struggling with navigation. However, commerce flowed on the sea so the navigators who made the best time were associated with the riches of the era. Nathaniel Bowditch perfected maritime navigation by combining his mathematical skills with his new nautical experience.

He is still considered the father of naval navigation. Every ship has a copy of his book, *The American Practical Navigator*, or *Bowditch* on their chart table.

PACT EXERCISE

As I share the Leadership PACT, you may think, *Why Admiral, this is just common sense.* I hear this often, and my response is always the same, "Most often, common sense is not that common." Usually, straightforward and simple advice is difficult to carry out or ignored altogether.

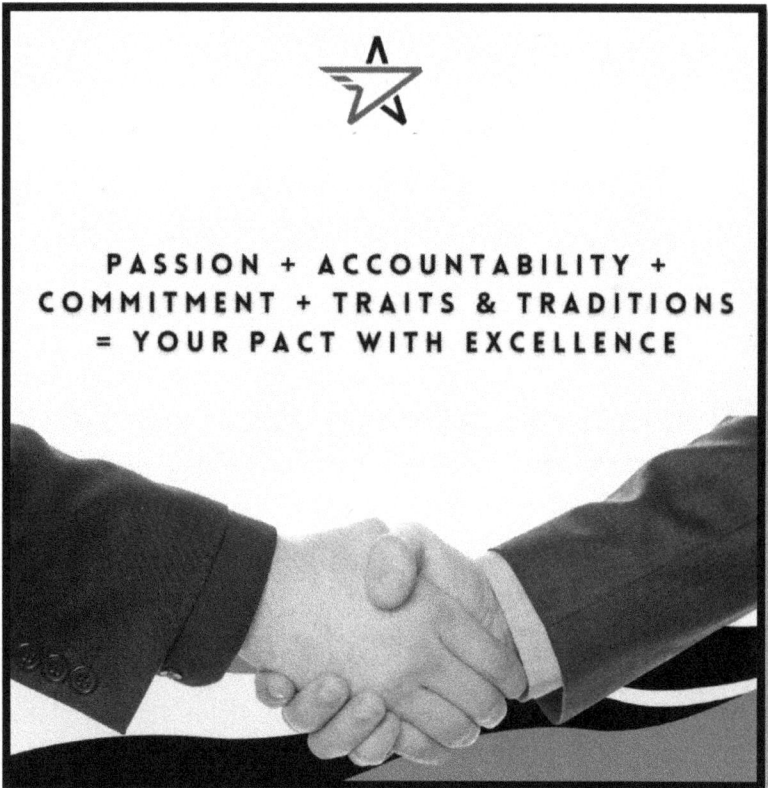

PASSION + ACCOUNTABILITY + COMMITMENT + TRAITS & TRADITIONS = YOUR PACT WITH EXCELLENCE

Let me repeat the PACT, and then we can take inventory of how ignoring this tool leads to failure.

"P" stands for passion and desire. Elite leaders desire to lead and display the passion required to lead.

"A" is for accountability. Elite leaders accept and take accountability for their teams. They singularly accept the consequences for their team's performance, celebrate their successes, and acknowledge the team. But, like any Kindergarten teacher will tell their class, "There is no I in team." Accountability stays with the leader, and only responsibility may be delegated.

"C" is for commitment. Elite leaders are all in when it comes to leading. Commitment guides and sustains you during the ups and downs, and successes of leadership. Not every day will be sunny and prosperous, but leaders must remain consistent and committed.

"T" is for the traits and traditions of elite leaders. As we navigate leadership, we learn from good and bad leaders and adapt traits and traditions to our personalities and style. Likewise, wise leaders learn from others and adjust as appropriate.

At this point, you should take a break and take inventory of your style and performance.

Now, take time to put pen to paper and try this exercise.

Write Passion, Accountability, Commitment, and Traits at the top of the paper. Under each heading, write short phrases, words, or bullets that come to mind concerning these attributes and your current leadership position. If you're adventurous, ask a close friend, colleague, or partner to evaluate your attributes.

Before starting this inventory with another confidant, have a conversation about the PACT so you both have an agreed understanding. This process takes courage but will reap tremendous rewards and provide you with a unique development map or navigational chart.

Navy Leaders run aground. What went wrong?

TAKING INVENTORY.

In the early 2000s, thirty-plus commanders of ships, submarines, and squadrons were relieved of command due to a loss of confidence in their ability to lead. This number of firings was double to previous years and set off alarm bells in the senior leadership of the Navy.

Moreover, the firings crossed over every demographic of race, color, creed, and gender. The Navy's head admiral wanted to get to the bottom of these firings and determine if a connection or trait led to qualified leaders being relieved of command.

While the number of leaders relieved was less than one percent of over 300 commands, the concern was to determine if there were systemic reasons for failure in command and to correct our course in training leaders. Military leadership is a high-stakes proposition because ineffective leadership puts national security at risk. Although your personal leadership situation may not carry the same weight, the same positive behaviors will make a tremendous difference for no dollar cost to your organization.

The Navy's Inspector General conducted a year-long investigation and found that there were three primary reasons individuals were relieved of leadership positions.

1. A significant event, such as a collision at sea, a ship grounding, or an aircraft mishap.
2. Organization failure to perform, such as not meeting combat readiness or failing a significant maintenance or engineering inspection.

3. Command climate, which is caused by adverse or ineffective leadership.

These categories capture broad behaviors of individuals and organizations, so a deeper dive into the investigation led the inspector general to come up with the acronym SEA, which stands for Sex Ego Alcohol. These three modifiers captured the behaviors that led to a leader's downfall.

Military leadership at its highest level can be intoxicating if one is not grounded, humble, and balanced. Suppose you obtain an important and influential leadership position. First realize that assuming this position does not make you funnier, better looking, sexier, or wiser. As was outlined earlier in this book, everyone is watching you and your personal behavior is under great scrutiny. If you become sold on your own importance and seek personal satisfaction over commitment to the people you lead, you will fail.

A leader who is not committed to the team and its mission tends to make everything about themselves. Their ego takes over, and the organization's climate descends into despair. Leaders with out-of-control egos abuse, manipulate, and take inappropriate privileges they don't deserve.

One great head of the Navy imbued us with servant leadership. We were reminded and understood that leadership isn't about the leader but the people we lead. The military axiom of *Take care of your people, and they will take care of you* never fails to make for a successful and high-performing team, no matter how small or how large.

Military leaders, more often than not, accept the consequences of their leadership. This is a sign of accountability, but still, there are failures that a PACT can prevent. Leaders without a PACT tend to lose commitment, accountability, and passion. They just give up, and those who do tend to resort to substance abuse, and in our study, that substance was alcohol.

Ethel Mae, my mother, always hammered home, "Sex, fire, and alcohol are not harmful when used in moderation." However, "All things in moderation," is common sense that isn't always easy to master.

THE 3 X 5 CARD TEST

One of my favorite Navy and Marine Corps leaders is retired Lieutenant General John Sattler. He is a true gentleman, warrior, scholar, and man of balance. In addition, he has a humor that lifts and lightens the environment no matter the severity of a situation. He's led in bloody combat, the halls of academia, and in the frantic offices of political and military leaders in Washington, D.C.

After retirement, General Sattler served as the Distinguished Chair of Leadership at the Stockdale Center for Ethical Leadership at the United States Naval Academy. He introduced the 3 X 5 card test with Midshipmen, and I've used it in several environments. It's another tool for quickly taking a personal inventory of your ethics and leadership. Here's the test:

1. Take a 3 X 5 card, and on the front of the card, write the name of a friend or colleague you can trust 100 % to have your back in a challenging situation. Someone you can call day or night for help and know that person will be there for you.
2. Underneath that name, write down several attributes that person has that confirm you can count on them 100% to be there for you. List as many traits as you can.
3. Hopefully, you feel pretty good and are smiling thinking about that individual. Now flip the card over and write down three names of individuals who, if taking this test, will write your name on the front of their card.

3 X 5 CARD TEST

· TAKE A 3 X 5 CARD, AND ON THE FRONT OF THE CARD, WRITE THE NAME OF A FRIEND OR COLLEAGUE YOU CAN TRUST 100 % TO HAVE YOUR BACK IN A CHALLENGING SITUATION. SOMEONE YOU CAN CALL DAY OR NIGHT FOR HELP AND KNOW THAT PERSON WILL BE THERE FOR YOU.

· UNDERNEATH THAT NAME, WRITE DOWN SEVERAL ATTRIBUTES THAT PERSON HAS THAT CONFIRM YOU CAN COUNT ON THEM 100% TO BE THERE FOR YOU. LIST AS MANY TRAITS AS YOU CAN.

· HOPEFULLY, YOU FEEL PRETTY GOOD AND ARE SMILING THINKING ABOUT THAT INDIVIDUAL. NOW FLIP THE CARD OVER AND WRITE DOWN THREE NAMES OF INDIVIDUALS WHO, IF TAKING THIS TEST, WILL WRITE YOUR NAME ON THE FRONT OF THEIR CARD.

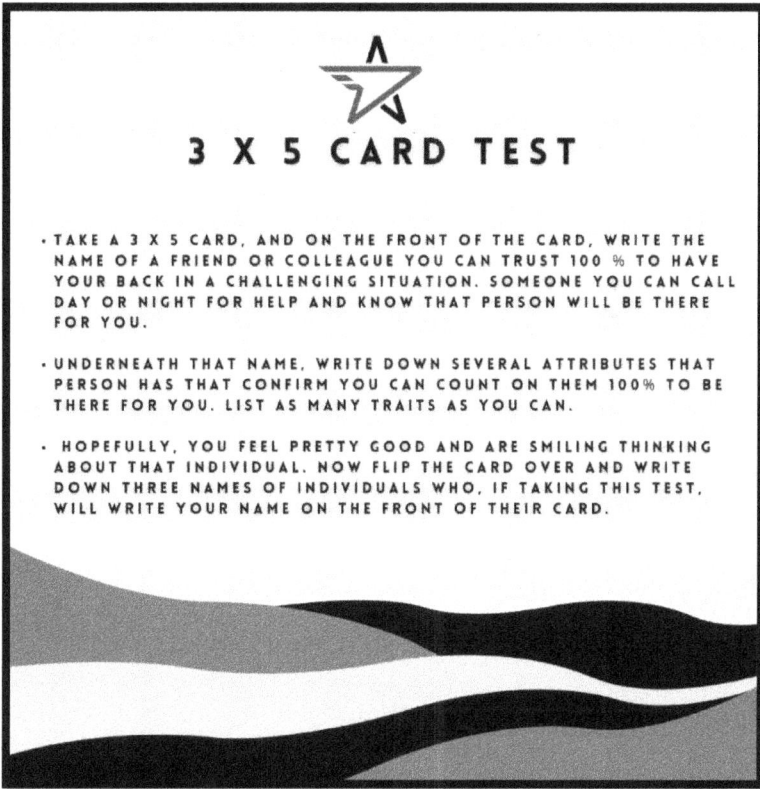

This is an eye-opener. If you have problems coming up with three names, flip your card over and reread the attributes you outlined. Suppose you aren't identified by those traits you listed. In that case, you need to examine why and rebuild your own inventory of leadership traits.

The two inventory tests in this chapter will give you a roadmap for your personal growth plan. This will aid you as you navigate elite leadership and success.

LEADERSHIP TAKEAWAYS

> ➤ Taking inventory is vital to your continued leadership growth.
>
> ➤ "Don't Give Up the Ship." Giving up is the first step to failing leadership.
>
> ➤ "All things in Moderation: Sex, Fire, Alcohol." *-- Ethel Mae Hall, My Mother*

ADMIRAL'S ADVICE

Two things.

➢ Start now! Grab some 3x5 cards, I use the cards, but you may want to use a notes app on your smartphone or tablet, and try General Sattler's examination of character

➢ Examine your life and behaviors. Ask a trusted advisor to help you with these questions. Are you about to run aground? Or do you lead with character? Be honest with yourself.

CHAPTER NINE

THE POWER OF THE PACT

> *"The quality of a person's life is in direct proportion*
> *their commitment to excellence, regardless of*
> *their chosen field or endeavor."*
> **Vince Lombardi**

The words *navigating* and *leading* both connote movement. If you navigate or travel to a destination, you find your way most efficiently. When leading, you're taking one or more individuals to a destination or an outcome. Whether you are headed for a family vacation, a business venture, or a military operation, you must start with the result in mind or, better stated, a vision of success.

In his book *Leading Change*, internationally renowned business author John Kotter states that step one in leading change is to create a sense of urgency. What better way to develop a sense of urgency than by establishing a compelling vision?

ELITE LEADERS CREATE A COMPELLING VISION

I first read Kotter's *Leading Change* at the Industrial College of the Armed Forces (ICAF) while I was the commandant at the National Defense University. Kotter's book helped me lead during a time of change.

The school's current name, The Dwight D. Eisenhower School of Resource Strategy and National Security, shortened to The Eisenhower School, is now more representative of the college. This new name is more appropriate than ICAF as General Eisenhower was a student at the school following World War I, he later taught at the school focusing on the logistic failures by the United States during the war. Further, the education Eisenhower received, and later taught, shaped his leadership as President of the United States during the development of a national highway system and warned of the military-industrial complex. The name change from ICAF to The Eisenhower School was met with initial resistance to the change and here is when Kotter's *Leading Change* methods came into play.

The compelling vision created was to brand the college, more appropriately, as The Eisenhower School. This name better describes our mission to produce leaders that would operate in the military and industry efficiently and effectively. Graduates of the college have gone on to the highest levels of leadership in the military, industry, and international and national politics. With 32 or more international students in each class, the school built a worldwide network of strategic partners that enhanced security across the globe.

The straightforward process outlined in Kotter's book was perfect for all leadership positions. As a result, I quickly became a disciple of *Leading Change*.

The Dean of our Leadership Department came to me with the great news that author John Kotter would be coming to the school to hold a seminar on leadership for our faculty during a school break.

Wow! I was excited to meet Kotter and discuss his book.

I was introduced to Mr. Kotter and praised his book, explaining how it had helped me. I thought his steps in leading change were the gospel of leadership.

Kotter was taken aback, and his body language showed he was uncomfortable with my effusive praise. He brushed off my comments, saying that his 8 Steps to leading change were just a simple idea he gathered over the years. He emphasized that his book wasn't the gospel on leadership, indicating that nothing in the book was written in stone, and the steps were certainly up for debate by any leader according to their personal leadership situation.

Since interpretation was up to each individual, the 8 Steps served as guideposts, and the ideal route could be individualized. His "Creating a sense of urgency" became my "Creating a compelling vision." Creating a compelling vision requires passion for the mission, in turn, building passion in your team. When leaders and navigators have a vision, they can see into the future, and they can communicate what they see.

MY FIRST MEANINGFUL AND INDEPENDENT LEADERSHIP OPPORTUNITY

As a mid-grade officer, I was assigned to my third squadron in San Diego, California, the Magicians of Helicopter Anti-Submarine Squadron Light 35. My first assignment in the Squadron was to be made officer-in-charge of Detachment 5. I would be assigned three junior pilots, one chief petty officer, the crew's senior enlisted member, two search-and-rescue aircrewmen who would fly with us, and twelve aviation maintenance petty officers. We were allocated one aircraft with all the tools, equipment, supplies, and manuals to support and fly the helicopter. Overcome with excitement, I wanted to prove myself as the best pilot in the Squadron.

Leading is all about people. Therefore, my first step in taking charge of this detachment was to learn about the people assigned to me and create a team vision. The Sailors I was assigned didn't care who I was or what I knew. They only wanted to see that I cared for them, their safety, their families, and their careers.

Using the crawl, walk, and run building process the team learned to operate and maintain our aircraft at sea on board a small Navy frigate or larger destroyer.

Recovering a helicopter to a small flight deck at night and moving it by hand into a hangar at night is the ultimate team-building event. Courage, strength, and brute force leadership by the more experienced Sailors are needed. After a month of practice and observation by a mentor, we were certified safe and ready to deploy on any Navy frigate or destroyer. We would soon be deployed overseas with the *USS Reasoner FF-1063* for almost eight months. Then, it was time to create a compelling vision for Detachment 5.

I watched the team come together and learned about their individual goals and aspirations, their family lives, and what motivated them. It was a ragtag group of individuals, but they were excellent as a team. I saw a great future in the deployment that was about to take place.

I spent hours thinking about a vision. I wrote by hand what I thought we could accomplish and how we would be recognized for our mission accomplishments and combat readiness. But I knew that, with my team, I could not just declare what my vision was. I needed their buy-in. During a break in the action, I had my team gather in the hangar. Once there, I explained that I would like to talk about our detachment and hear their thoughts on what they would like to accomplish operationally during our upcoming 7-month deployment. I wanted their vision even though I had written out a clear vision myself.

Their responses were unanimous. They desired to be the best detachment. Whenever I hear a modifier like "the best," I question what this means and what it requires.

The process continued as I asked more questions. We teased out what it meant to be the best, how much work was required, and how many maintenance and flight evolutions would be necessary.

Although the wording was slightly different, the team came up with all of the actions I had written down. Most importantly, they had bought into the compelling vision of being the best helicopter detachment in the Navy. They knew the hard work required and were enthusiastic about accomplishing their goals. My vision became their vision. A total buy-in.

A leader knows that a goal or vision without a timeline and a plan is only a dream and not a vision. I worked with the seasoned team members to create an eight-month calendar that detailed our missions, flight time, deck landings, and required maintenance per flight hour. Keeping in mind that every team member needs balance, we also calculated time off for recreation and phone calls home to family and friends.

At this point, we had pulled together a team built on the basic requirements of operating a helicopter detachment and had qualified in our warfare requirements of anti-submarine and anti-surface warfare. We had a vision, a plan of action, and milestones to reach our vision. The next step was to communicate the vision.

In the early 1960s, President Kennedy declared we would put a man on the moon and return him safely to earth. This was a powerful and compelling vision. It was now up to NASA to devise a plan of action and provide milestones that would fulfill the President's foresight. For this bold mission, buy-in by everyone involved was imperative and communication at every level of the organization was critical.

Legend has it that visitors to Cape Canaveral, the rocket launching sight, asked a custodian sweeping the hangar floor what his job was.

His incredible response was, "My job is to assist in putting a man on the moon and safely return him to earth in this decade."

In the case of our helicopter detachment, everyone in the chain of command needed to communicate our vision. The squadron command leadership and the ship's captain all needed to understand our vision, goals, plans and resources. As a result, we had a collective idea of what being the best helicopter detachment entailed.

A vision without a timeline, though, is just a dream. Therefore, over the next twelve months, we worked as a team, remained committed to our vision, executed our timeline, and hit all milestones. Of course, there were bumps in the road, but we could adapt and continue to execute because we had a PACT with our mission and a detailed plan.

When the deployment ended, the Secretary of the Navy recognized our ship and helicopter detachment team as the best in the Navy. All of my pilots and senior petty officers were awarded Navy Achievement medals. In addition, all junior personnel were given Admiral level letters of commendation, adding points to their promotion process.

Later that year, my team submitted me for recognition. I was selected as the Navy's Helicopter Pilot of the Year for sustained performance. What a year of adventure, deployment, flying, and contributing to our Navy's combat readiness. My leadership path was coming together in the most positive way.

GOAL SETTING TO NAVIGATE SUCCESS

Early in my career, I became fascinated with non-fiction books focused on success. The master salesman Zig Zigler hit home with me in his book, *See You at the Top*. I loved his phrases like "eliminate stinking thinking" and "get a check-up from the neck up."

After reading his book, I started responding to the question, "How are you?" "How are you doing?" and its many variations with my own. "I'm doing super-duper, but don't worry about me because I'll get better." I've been saying that for over four decades to friends and strangers. It's fun to see people's responses which often include a smile, a quizzical look, and the question, "What's better than super-duper?"

This interaction may seem rote and repetitive, but it keeps the mind positive. If you are a leader, it pays to have a positive outlook even when the chips are down. Being positive is not mutually exclusive to being serious. Positivity is a force multiplier.

One night on my helicopter detachment deployment, the ship was awakened with the call to launch the helicopter at 2:00am in the middle of the night. An unknown submarine was in the area, and we were needed in the air to detect and localize it as it wasn't responding as a United States submarine should. Therefore, this was an unplanned evolution.

I hopped out of my top bunk, slipped on my flight suit, and headed to the combat information center to get briefed on the situation. On the flight deck, my crew was prepping the helicopter for launch. They were a well-oiled team and worked flawlessly even though they'd been startled awake. My co-pilot ensured the bird had the proper ordnance and was pre-flighted properly.

Heading up to the flight deck, I cut through the officer's dining room. The bright lights were on, and the Captain and the Commodore sat gulping Navy black coffee.

The Captain whispered to the Commodore, "Watch this." Then, turned to me and said, "Hall, how are you feeling?"

Without hesitation, I responded, "**Super Duper,** Captain. But don't worry. I'll get better."

Before heading to the flight deck, the Captain told the Commodore, "It's crazy. He's always like that."

I am of the opinion that if you are going to take off from a frigate at sea in the middle of the night, you better feel super-duper, and your crew needs to know this.

I bounded across the flight deck, swung into the bird named *Sea Sprite (SH-2F)*, and strapped myself to the helicopter that had become my extra appendage. I was not simply a pilot; I was one with a powerful machine.

My exit off the aircraft was met with smiles and several thumbs up. It was time to hunt submarines. I was living a real-life video game. I was super-duper, and everyone knew it. I still hear from that crew from time to time, and when I ask how they are doing, they respond, "super-duper."

Keep in mind, no one enjoys following a sour puss. A leader who inspires has a goal, a compelling vision, a detailed plan, and builds a positive environment. And a sense of humor never hurts.

Zig Zigler gave me that positive outlook and its power over others. My reading continued with books like *The Seeds of Greatness* by fellow Naval Academy graduate Dennis Waitley and *Think and Grow Rich* by Napoleon Hill.

It was Napoleon Hill who gave me the inspiration for the process of goal setting. I had set goals, achieving some and falling short on others, but Hill's process was something I understood and could tweak to my situation. Hill's goal setting is focused on money and military currency is the mission or combat readiness.

THE PROCESS OF THINKING AND GROWING RICH GOAL SETTING

All one needs to do is enter *goal setting* in a search engine, and pages-upon-pages of results will be returned. The process described in *Think and Grow Rich* will show up as one of the top results of your search. The process is straightforward and relies on manifesting your goals. Hill's famous saying is, "If you can conceive it and believe it, you can achieve it."

His process is a six-step process. Before I give you my interpretation, I want to tell you about putting this process into action. Intuitively, I knew that when writing out goals and plans, the best method involved pen on paper. You may think, "Admiral, come into this century. No one writes like that. We use tablets, smartphones, and keyboards." However, putting pen to paper engages the brain in multiple modalities.

While serving in the White House, I worked with over a dozen lawyers in various roles, not all in legal areas. I noticed that they took the best meeting notes, using pen and paper. When I asked about their notetaking, each said it was a technique learned in law school and that you use all of your senses with this old-fashioned method.

The brain, muscles, pen, and paper all worked together to lock in the learning process. When writing your goals by hand, you're engaging more of your brain and imprinting your goals to memory. Thus, you are more committed to carrying the process out and achieving your goals. Not only this, but you have the additional benefit of improving creativity.

THE METHOD

Step one: Be specific in your goals. The more detailed your goal statement, the less wiggle room is involved. For example, saying, "Our goal is to be the best helicopter detachment," does not work. Who or what determines the definition of *the best*? The more specific you are in describing your goal, the less room is left for excuses. A psychological advantage of being specific is that your mind can visualize the final outcome.

Step two: State what actions you're willing to do to achieve your goal. Your personal effort is needed, and you need to know what you're ready to do and what you can do. Writing this out by hand is imperative as you commit to yourself. This will also help you take inventory of your personal skills and resources available. After taking stock of what you are willing and able to do, it is the perfect time to look back on step one and see if you have described and set the proper goals.

Step three: Set a time and date for achieving your goal. You can wish for success every day, but without a timeline and a completion date, it is just a wish. There are two possibilities for setting a timeframe for success. One, you may have been given a deadline by your superiors, or two, you have the liberty to set your own timeline. In each case, you can reevaluate your goals and inventory of resources. When picking your own time for success, consider discussing the process with a mentor who has been down a similar path. Choosing too short of a timeline can induce unneeded stress. Choosing too long of a timeline may cause a haphazard approach to achieving your goals. It's a Goldilocks situation. You need the "just right" timeline.

Step four: Write out a clear and detailed plan. In the military, we often refer to this as a POAM or Plan of Action and Milestones. This is where you create a calendar and place all the intermediate steps leading up to your goal. If you put the right amount of thought and planning into this process, I guarantee there will be a step to take today. Every journey begins with one step, and your path to your goals starts today. This plan is intended to be a living document, so leave space to add steps and corrections as you move toward your goals.

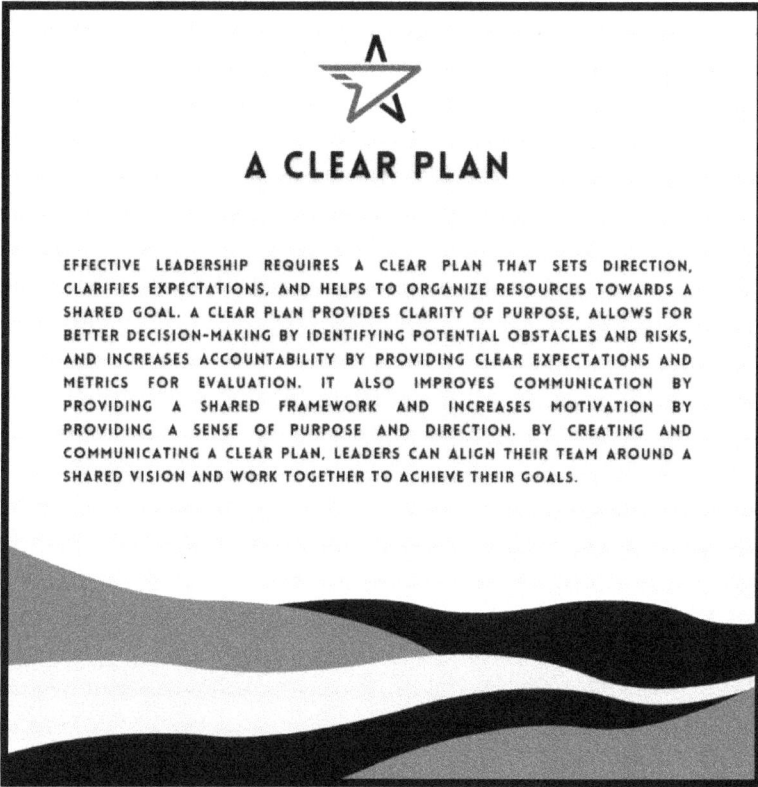

A CLEAR PLAN

EFFECTIVE LEADERSHIP REQUIRES A CLEAR PLAN THAT SETS DIRECTION, CLARIFIES EXPECTATIONS, AND HELPS TO ORGANIZE RESOURCES TOWARDS A SHARED GOAL. A CLEAR PLAN PROVIDES CLARITY OF PURPOSE, ALLOWS FOR BETTER DECISION-MAKING BY IDENTIFYING POTENTIAL OBSTACLES AND RISKS, AND INCREASES ACCOUNTABILITY BY PROVIDING CLEAR EXPECTATIONS AND METRICS FOR EVALUATION. IT ALSO IMPROVES COMMUNICATION BY PROVIDING A SHARED FRAMEWORK AND INCREASES MOTIVATION BY PROVIDING A SENSE OF PURPOSE AND DIRECTION. BY CREATING AND COMMUNICATING A CLEAR PLAN, LEADERS CAN ALIGN THEIR TEAM AROUND A SHARED VISION AND WORK TOGETHER TO ACHIEVE THEIR GOALS.

Along with a POAM, Naval Aviators have an OODA Loop, which is most common among fighter pilots as they encounter an enemy aircraft. But first, I pose the question, "Do you know how to tell if someone is a fighter pilot?" The answer. "Don't worry. They'll tell you?"

The OODA Loop stands for the process: orient, observe, decide, and act. Repeat. Orient your plan, and then observe how it works.

Make decisions based on your observations and take actions as appropriate. Repeat this process over and over as you refine your plan and zero in on your goals.

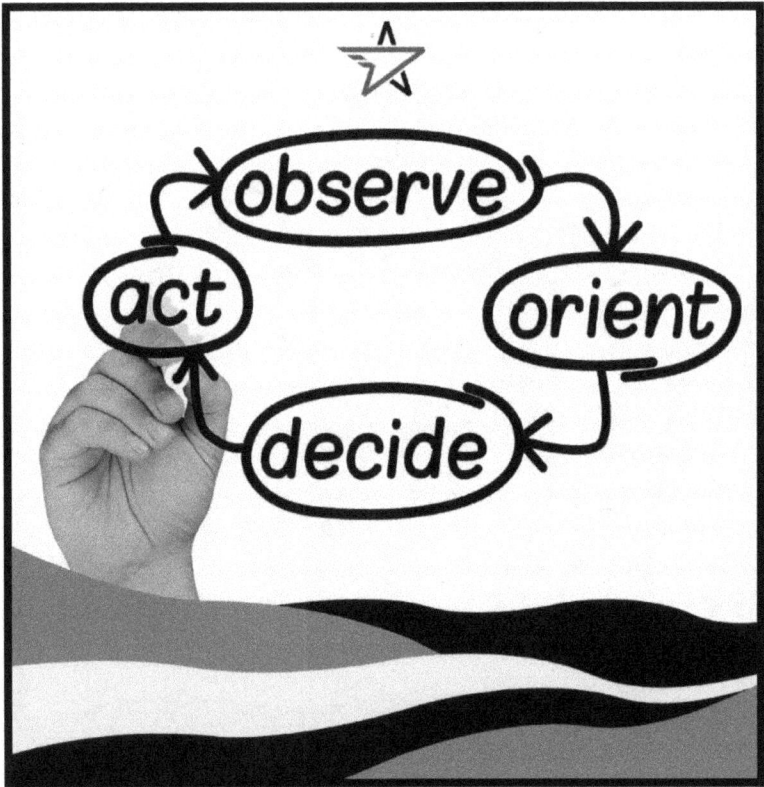

Step five: Write out a concise statement. Once you have your goal set and your plan of action, and how you will contribute, hand-write a brief summary. Here, the brain, hand, pen, and paper work together to solidify your goals. This statement of your intent, attainment date, and what you are willing to do to achieve the goal should fit on a 3 X 5 card. Keep working on this statement and share it with friends and mentors until it captures your intent and fits on that card. By this step, you are becoming laser-focused on your success and are assured of obtaining your goals.

Step six: The final step is creating your positive affirmations. At this point, you've set a great goal by establishing a timeline. It's no longer a wish or a dream. You have a plan and a concise statement of the path to your goals. You have a step that can be taken today. Ready, set, go! Read your brief statement out loud twice a day. Once in the morning to start your day and the second time before retiring for the night. You'll be programming yourself for success and for problem-solving.

These six steps pull together academic, scientific, and physiologic processes that have led to success for over a century in businesses of all kinds. As leaders, goals are crucial because you guide a group to a destination or on a journey. Once again, balancing your life is a requirement for success. Therefore, I recommend you have not only professional goals, but personal and family goals as well.

LEADERSHIP TAKEAWAYS

> ➢ Leaders create a compelling vision and elicit excitement and engagement to follow that vision.
>
> ➢ Communicate your vision throughout your team in writing spoken words and actions.
>
> ➢ To lead is to set goals and create a detailed plan you start today.

ADMIRAL'S ADVICE

Start now! Break out those 3 by 5 cards again. I use the 3 by 5s but you may want to use your smartphone or tablet. Now write out your goals with a timeline. Start with a short-term goal that is reachable and see how quickly you achieve and surpass your goal. You will now have the confidence in the process to go after more lofty goals for yourself and for your team.

CHAPTER TEN

PUTTING THE PACT INTO ACTION

Understanding the PACT is simple on face value. The electrifying part is when you put the PACT into action. Holding space for up-and-coming leaders excites me, and as I engage in conversations with them, I love to see their eyes light up when they think about their future. The possibilities are endless. The first step is to gain clarity with your personalized PACT.

These questions will help you dig deeper into the *Why* behind your vision and propel you to move forward even when met with resistance. Unfortunately, we often become our biggest distractors. This can look like imposter syndrome, self-sabotage, or lack of confidence. What is important to remember is that we are all human beings. We all have the potential to be at the top of our game. Still, we often get in our own way.

Walking through an exercise of answering the questions is a lot like writing a mission statement or business plan. When we articulate our goals and explain the methods to obtain them, we have the formula that keeps us on course. And we cannot do this alone. It is nearly impossible to reach our goals without the underpinnings of a supportive team.

Looking back at one of the best football players of all time, Quarterback Tom Brady, we can see how he epitomizes excellence. Brady also exemplifies the importance of surrounding yourself with a fabulous team. Brady would not have seven championship rings without his coach, mentor Bill Belichick, and his two exciting and talented receivers, Randy Moss and the colorful Rob Gronkowski.

A little-known fact about the Belichick family and legacy is that Bill's father, Steve Belichick, served in the Navy during World War II. He played in the NFL and then coached at the college level. He spent 34 years as a scouting coach at the United States Naval Academy. He is considered the expert on college football scouting, and he left the Naval Academy with the most extensive library of football books outside of the NFL Hall of Fame. To this day, Bill continues to add to this library.

Like Tom Brady, you can be good. You can be really good at anything–maybe even the greatest of all time. Yet, if you want to win, you need a team. By leveraging the strengths of others, you have the power to brainstorm. Enlisting the support of a devil's advocate, you can learn from another's vantage point. Our brains are not created to know it all; it is about leaning on others and lifting each other up. The power of a team will help you cross the finish line.

Remember, putting pen to paper helps clarify your goals and creates a roadmap to your overall success. Answering these questions will lead to a clear vision and blueprint to follow.

CREATE YOUR PACT

Let's take a break, put down this book and pick up your pen or pencil. Grab your pad or notebook, and let's write your PACT.

You may not be able to answer all these questions in one sitting. I encourage you to give it a shot and then discuss the assignments with a mentor or friend who has your best interests at heart. Sometimes I do this with a longtime mentor and classmate. However, most of the time it is Darlene who assists me. I don't always like others' input, but I always take time to consider it. More often than not, they're right because they know me and my style and potential.

P - UNDERSTANDING THE
"WHY BEHIND YOUR LEADERSHIP"

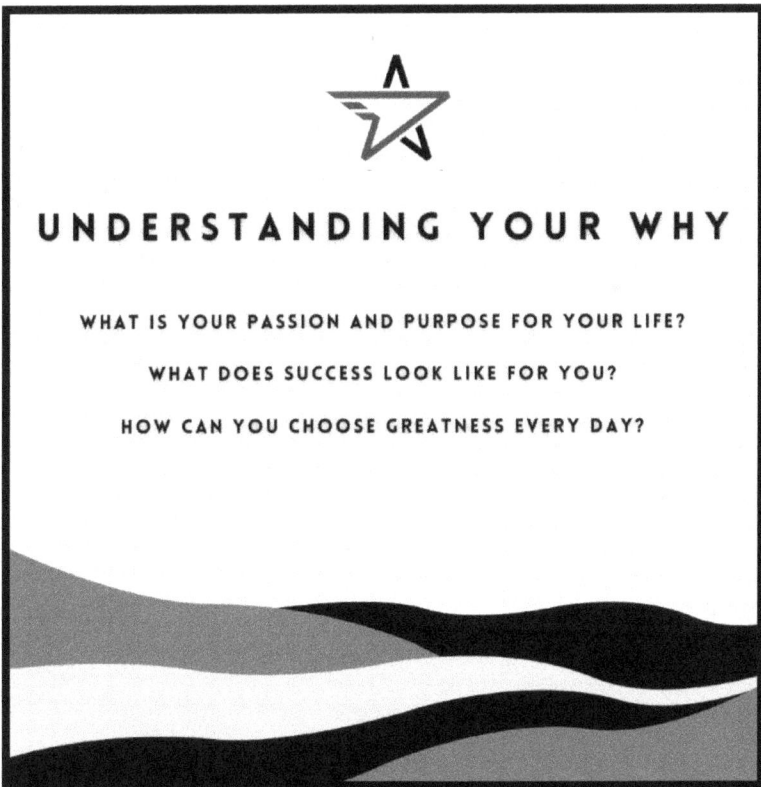

1. What is your passion and purpose for your life?
2. What does success look like for you?
3. How can you choose greatness every day?

A - Staying Firm to Your Vision

1. What actions can you take to ensure you remain accountable to your vision?
2. Who can you recruit as an accountability partner to ensure overall success?

C - Maintaining the Anchor of Your Vision

1. How can you serve your vision as a leader and commit to the success of your plan?
2. How can you communicate this vision to others?
3. Are you committed to living out your values every day?
4. How can you maintain quality control when navigating rough waters?

T - Strengthening Your Mission

1. How will you implement your leadership strategy that aligns with your mission?
2. What resources will you use to maintain your plan of attack?
3. Who can you reach out to as a mentor?
4. How will you build your team?

FINAL QUESTION:

What's holding you back from creating this pact with leadership?

It may look like imposter syndrome, fear, or an inflated ego. But what's handcuffing you from leading an extraordinary life? We all have choices in life. We have large, small, and important decisions to make. I've learned that making the right small choices adds up and leads you in the proper direction, so you never have to make an important and challenging big choice.

You can choose to be average, over-indulge in life, and surround yourself with people who don't share your values. So, take the step to select greatness in all things, small and large. I guarantee you that you can make a small choice or decision today that will propel you to be an elite leader tomorrow and well into the future.

TAKE 5: A CASE STUDY, DOES LEADERSHIP WORK?

I previously stated that leadership will improve your bottom line without adding costs to your operation. You need to decide what your bottom line is. Is it readiness, like so many cases in the military, or is it hard cold cash earned by an organization?

In my local neighborhood, there is a maintenance and oil change franchise.

> *"We're all about fast, friendly, and simplified oil changes. We change your engine oil along with the filter in just a few minutes. Vital under-hood fluids are checked and replenished to the recommended level."*

Additionally, they have earned several customer satisfaction awards. Every franchise, whether a hamburger joint or an oil change operation, has its uniform set of rules, infrastructure, operating guidance and recipes for success. Why, then, do you enter some franchise restaurants and your feet stick to the floor, or are you met with indifference?

LEADERSHIP MAKES A DIFFERENCE

In my area of Hill Country, Texas, two of these oil change franchises are within ten miles of each other. Both are in neighborhoods with similar populations and demographics and thus should have similar service demands. The location that was recommended to me by a friend five years ago continues to service my cars and trucks with oil and filter changes and state vehicle inspections. They do a booming business. Ten miles away, the other franchise has a larger layout and more service bays for oil changes but does a fraction of the business of the franchise I frequent. *Why?*

A look at the Nextdoor website and Yelp reviews provide the answers. The underachieving franchise is reported to have shabby workmanship, filters improperly installed, after-service leaks, and disgruntled employees. The current manager is often found sleeping in his office, taking longer than usual lunch breaks, and often berates his employees in front of customers. He is also known to argue with customers and scold them if they turn down recommended additional services. All this activity gets documented in surveys and online reviews for all to see.

The franchise I utilize has two men of maturity that run the shop. Both are easily identified as gearheads, which means individuals who love working on cars. Their calloused hands and broken fingernails quickly show how hardworking they are. In addition, they have **Passion** for their work.

The store leaders take their job seriously and can be observed working closely with their employees, all young men and women. I have watched them use clear and concise communication when working as a team servicing my car. I equate it to the communications religiously used in multi-piloted aircraft, where communication is a number one safety factor. The managers and crew are **Accountable** for the service they provide and to each other. The two seasoned managers communicate to their employees in a friendly but professional manner; they display their **Commitment** to the mission of vehicle service and to growing their employees in the workforce. Their young employees are generally new to the workforce. They're learning **the Traits** and **Traditions of** quality work and customer service.

The operators of this franchise have intuitively made a PACT with leadership, and it has served them well. As a result, they have great reviews on Yelp, increased business, and a bigger bottom line. Additionally, they are giving a great start to young men and women who earn a living wage and navigate the workforce.

Navigating leadership is for everyone, not just executives and entrepreneurs. Your PACT with leadership will positively impact your personal and professional life.

LEADERSHIP TAKEAWAYS

> ➤ Leaders write. Writing helps you collect your thoughts concisely and programs your actions.

> ➤ Build a team around you that compliments your style but is willing to give you input and challenge you. Never surround yourself with only "Yes" *people*.

> ➤ Leadership makes a difference in the bottom line of your currency.

ADMIRAL'S ADVICE

Take time and answer this chapter's thought-provoking questions. Ask yourself, what does success look like in six months to a year. Answer these same questions again at six- and twelve-months intervals and make corrections to your PACT as needed.

CHAPTER ELEVEN

THE LEADERSHIP BOTTOM LINE

The bottom line of leadership is to increase currency without increasing costs. What is your currency? If you are in business, your currency is dollars added to your bottom line. Why is that? First, employees do not leave companies, corporations, or large organizations. Instead, employees leave their first-line supervisor. Quality leadership retains employees, and retention saves money in finding, hiring, and training new employees.

Next, quality leadership creates an environment where an organization operates efficiently and effectively. Effective and efficient teams simultaneously save money and earn money. A win-win business situation. Leadership matters.

Too often, an organization motors on without leadership, and employees quietly quit, meaning they come to work, and they do the bare minimum to stay employed and no more. Unfortunately, we are seeing this phenomenon more and more in the post-pandemic work world.

Leadership in a for-profit organization makes sense to nurture, empower, and develop. Every dollar invested in improving leadership is returned in multitudes.

Military organizations have different currencies such as budget dollars, resources, training, maintenance, and/or combat readiness.

As a result, they invest heavily in leadership training and development. The industry is slowly realizing this as they actively recruit and hire people who've served, regardless of their service time. It needs leaders, innovators, and problem solvers.

When I was captain of *Tarawa,* I considered one of my roles to be retaining qualified and quality Sailors. But if they were determined to serve only one four-year enlistment, I would thank them for serving and happily help them with their transition.

Transition is something the services take seriously. Often four years in the Navy is more beneficial than four years in college. One of my Sailors served a four-year enlistment entirely on *Tarawa*. He developed his leadership skills, achieved the Petty Officer Third Class rank, and earned his aviation warfare badge. He took every advantage the Navy offered to develop his leadership skills. He made his own PACT with leadership. When I asked him to stay in the Navy, he said he was offered a civilian sector offer he couldn't refuse.

When he shared his job offer, I agreed that he served with honor and was ready to move on to greater opportunities. A large footwear corporation hired him at an income just shy of six figures, offered him and his family full medical and dental coverage, paid for his household move, and placed him in their management and leadership development program. His leadership in uniform enhanced *Tarawa's* combat readiness and now would positively impact this new company's bottom line. Leadership matters.

If you feel you are too junior, inexperienced, or young to lead, I ask you this question. Are you on a team that includes you and at least one other person? If yes, you can start your own PACT with leadership and begin to lead. *How do I do that*, you ask? Do the following, and I guarantee your success in any organization.

1. Execute on your PACT and lead by example.
2. Lead by showing your passion for your position by taking full accountability for your actions and responsibilities.
3. Show them your commitment to the team's mission.
4. Develop the traits and traditions of quality leaders you have observed.

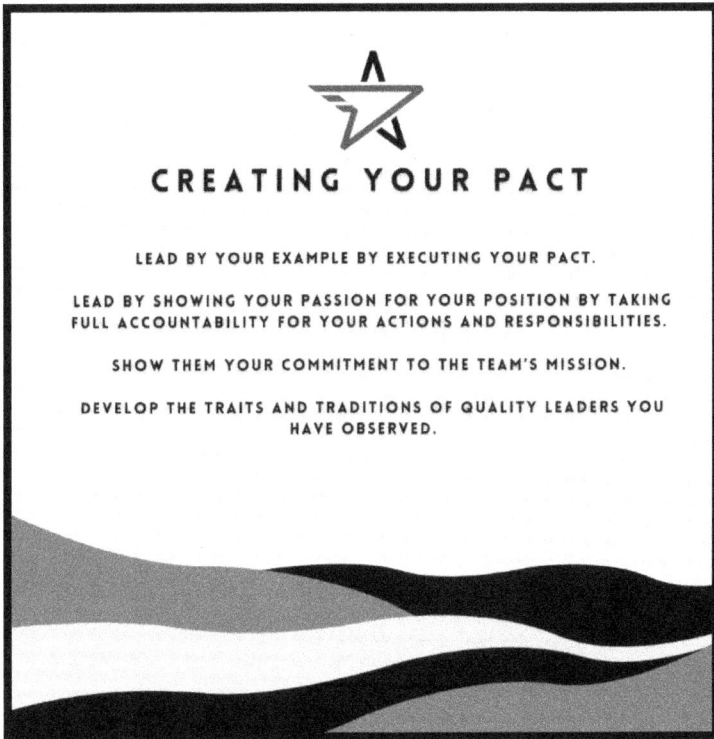

CREATING YOUR PACT

LEAD BY YOUR EXAMPLE BY EXECUTING YOUR PACT.

LEAD BY SHOWING YOUR PASSION FOR YOUR POSITION BY TAKING FULL ACCOUNTABILITY FOR YOUR ACTIONS AND RESPONSIBILITIES.

SHOW THEM YOUR COMMITMENT TO THE TEAM'S MISSION.

DEVELOP THE TRAITS AND TRADITIONS OF QUALITY LEADERS YOU HAVE OBSERVED.

Having commanded two squadrons and a capital ship, I was always happy to meet with a Sailor's family. The men in my family have had premature gray hair at an early age. A business executive visiting the *Tarawa* once asked why as a young captain, I didn't color my hair?

"The parents of my Sailors who put their lives on the line for their ship and shipmate want to see a captain with gray hair. They want to know someone with great experience is leading their child," I explained.

Parents would often ask me what advice I have for their children to be successful? My answer. "I believe there are four steps for any young emerging leader in any organization to succeed."

1. Show up to work on time, rested, and ready to go. Staying up to all hours playing video games or binge-watching videos is not conducive to success.
2. Show up to work clean in mind, body, and uniform. No one wants to work with someone who smells of body odor or cigarette smoke. No one appreciates a colleague who is in a wrinkled, slovenly outfit. You cannot be an effective employee if your mind is in the gutter.
3. Do more than is expected and without being asked. You know what to do, so do it.
4. Do your work with a smile. Your positive personal outlook is a force multiplier and lifts your spirits and the spirits of those you work with.

Succeed in doing these things, and you'll be identified as a leader and make a difference in your organization and your personal life.

LEADERSHIP MATTERS

Warning: In today's post-pandemic world, where the work-from-home option is widespread, it is easy to only manage and not to lead. Do not be the leader who develops keyboard courage and neglects taking care of people in your organization and your team.

Keyboard courage is the act of hiding behind the veil of the computer and saying and doing things you would never do in a face-to-face situation. Keyboard courage leads to unkind remarks, misunderstood communications, and in the worst case, layoffs without face-to-face interaction. Video calls are essential but shouldn't be your only method of connecting with your team. The fundamental attribute of the PACT with leadership remains. Leadership is connecting with individuals in a meaningful way.

LEADERSHIP TAKEAWAYS

> ➤ You will have failures and frustrations. However, you're not judged by your fall but by how fast you get up.
>
> ➤ Take the time to get your priorities in line with your organization. But, more importantly, if you are in a relationship, get your priorities right and in alignment with your partner.
>
> ➤ Listen more than you speak.

ADMIRAL'S ADVICE

The best advice I've ever received in our high-tech smartphone and device world is this: When you are physically present with someone, be mentally present. How often parents or spouses are physically present but mentally elsewhere. Put down the phone and the tablet and focus on those around you. It is easier to say than do, especially when you know you have tasks to complete, emails to respond to, and projects on a deadline.

CHAPTER TWELVE

THE PRICE OF LEADERSHIP

The onus of this book is to communicate that with every choice we make as a leader, we have to sacrifice something. In the first few chapters, you may have had the sense that I was an arrogant son of a bitch. This was intentional because I was.

As I ascended to higher positions of authority, I made the unconscious (or maybe conscious) decision to put my ego first. In my gut, I knew I was sacrificing my family. And to be honest, a distant, dysfunctional family is all I knew. I chose to forgo evening meals with Darlene and the kids to serve a different purpose. I decided to miss Courtney's recitals instead of a board meeting or a night out with the guys. Unfortunately, I came home to a woman in despair. At my core, I wanted to be loved. I yearned for adulation and applause. And I was blinded to see that everything I needed was within the four walls of my home with my wife. My loved ones paid the price. They followed me from post to post, job to job and were uprooted from their friends and security.

My priorities adjusted when I got the proverbial slap in the face I needed. I applied the Leadership Pact with the firm understanding that my family came before my job. And this is when everything changed. I hope that by sharing my missteps, I have given you a lens to examine your life path.

Understanding your passion and priorities is number one. Number two is anchoring yourself with a team who holds you accountable to this mission. Finally, following up with clarity of your commitment, rooted in the traditions of those that came before you, will ensure your journey to the top.

It is not easy to unlock this potential, but it is possible. I, most certainly, am an unlikely leader. Sixty years ago, my third-grade teacher believed in my potential. Still, I am unsure that anyone else had faith in me. Understand that with every failure, you can learn life's greatest lessons. If I had not hit a wall of despair, I would be on the same path, wheeling and dealing, and probably divorced. Instead, every failure is a building block towards a better version of ourselves. Every bump in the road allows us to hold on to loss or learn, then pivot.

Pivoting is the secret to true success. When you are working with a team, whether it be with board members or your spouse, feedback is essential. Listening to your trusted advisors is a skill that cannot be avoided. When we eschew feedback, we fail to understand our situation from a 360-degree vantage point.

Listening skills separate greatness from mediocrity. Have you ever talked with someone who seems to answer before you've completed your sentence? In essence, they are formulating their response and not effectively listening.

Effective listening leads to effective leadership. This skill does not come easily for a lot of people. However, when mastered, you can become a person of influence. Studying body language and tonality and delving deeper into the words used unveils the power to connect, understand and make judicious decisions. Effective listening is a tool that must be learned, then practiced daily.

ACCELERATING YOUR LISTENING SKILLS

You have two ears and one mouth, use them proportionally. In other words, listen twice as much as you talk.

Don't prejudge the person you need to listen to. It is human nature to judge someone when they walk into a room or jump on a zoom call. Our initial response when meeting someone is to scan their appearance and make a judgment call, which can lead to a negative filter in which we absorb their words. It's essential to actively choose not to prejudge and listen to the conversation's inflections, tone, and context.

Make every attempt to refrain from spouting a reply before the other person completes their thought. This requires discipline and being present in the conversation. This will allow you to understand what the other person is trying to convey and help you to respond appropriately.

Listen for the facts. Communication is expressed primarily through words, tone, and body language, and experts say that seven percent of a message is conveyed by words. Therefore, dissecting the conversation and pinpointing the hard facts will allow you to respond precisely.

Avoid Distractions. With the advent of technology, everyone has their cell phone glued to their hip. When we are multitasking or distracted by texts, social media, or checking out the score of our favorite football team, our conversations are diluted or lost. Make every effort to minimize distractions. Put your phone on silent, turn off notifications, and connect with the person in front of you.

Hindsight is 20/20. Now I realize it was more important for me to be heard in my youth. After all, all humans have an innate need to be heard and seen. Unfortunately, at the onset of my career, I believed that the loudest one in the room won. I was not yet mature enough to understand that the one who listens has the true power.

LEADERSHIP TAKEAWAYS

> ➢ Always be present – listen with eye to eye contact and put down your devices.
>
> ➢ Listen more than you speak.
>
> ➢ Start these steps now. As Chaplain Don den Dulk would day, "don't should on yourself." Do not look back on previous behaviors. Move forward today with these tips.

ADMIRAL'S ADVICE

Take time to listen to subordinates, regardless of their status or experience and process what they are saying to you. A Navy ship had a collision at sea that could have been avoided. All the experts were on the bridge assessing the situation leading up to the collision. Only one person identified the impending collision, a young Sailor. She had been a trouble maker and marginal performer so she was ignored by the officers in charge. The following investigation cited her observations and attempt to be heard would have prevented the collision. Listening is powerful.

CHAPTER THIRTEEN

CHOOSING GREATNESS

"You were designed for accomplishment, engineered for success, and endowed with the seeds of greatness. Greatness is a choice."

Zig Ziglar

Navigation is a process of getting from point A to point B. The most crucial point of navigation is determining where point A is located by determining your geographical position or location. In this book, we are not concerned with your geographical position but rather your personal and professional location or situation.

We often use land navigation when we leave our homes. But, most likely, if you are like Darlene and me, you use the most current application on your smartphone or car's navigation system. It is funny how reliant we have become on GPS apps to get from our houses to our destinations.

THESE GPS APPS ARE SIMPLE TO USE AND ASK SO LITTLE

After you turn on your GPS, you are prompted with a couple of options or questions. *Do you wish to go now? Depart from your current location? What arrival time is desired? So many choices.*

When navigating at sea, you have many influences; from sea state, currents, wind and weather, and ship capabilities to fuel state. To complicate things, your destination may move, such as a refueling ship you are scheduled to rendezvous with. The captain and navigator working together have multiple options to consider making their assigned replenishment time and location. *Oh, those choices.*

Mathematically, the slightest deviation in your early course has the exponential effect of putting your ship miles off course requiring a significant change to recover. For example, a one-degree course error at sea can put you sixty miles off your travel plans. Likewise, placing you a mile from your target requires a ninety-degree adjustment to recover. So, that one small choice can change everything.

THE CHOICE EQUATION

Choices + Errors + Acknowledging Errors= Rerouting

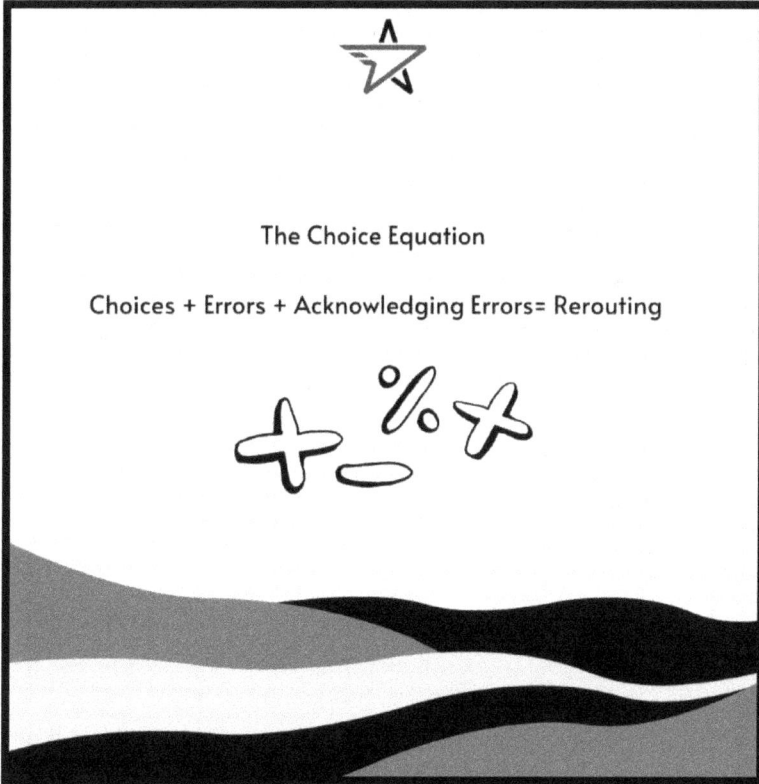

The Choice Equation

Choices + Errors + Acknowledging Errors= Rerouting

In this scenario, a minor course correction early in your journey will prevent a significant course correction along your travels. I hope you recognize these deviations from your mission and attempt to correct them. After all, our failures teach us how to be better humans. No one is perfect, as you can see from my experiences. However, we gain clarity if we objectively look at our paths, taking the emotional response away.

Earlier in this book, when discussing our marriage, I said I make all the big decisions, and Darlene makes all the little ones. And to show you how good I am, I haven't had to make one decision in forty-five years. While this is an amusing story, Darlene has a history of making small, well-thought-out decisions or choices. We've never needed to make radical choices to get back on track with our lives. *Again, choices.*

Are there big decisions and choices to be made in everyone's life? All of the time, and these meaningful choices are made more accessible when you make good decisions earlier in your personal and professional life.

SEA STORY

On the flight deck of a Navy carrier, the noise level is so loud every Sailor wears hearing protection and then communicates with hand signals. When an emergency arises, we resort to a loud whistle that stops all movement. Each Sailor wears a safety whistle around their neck and is ready to use it with the speed of a skilled basketball referee because lives depend on quick action.

When it isn't around my neck, my whistle dangles from my keychain. One of my colleagues likes to quote me. "I have a whistle, and I'm not afraid to use it."

As captain of the *USS Tarawa*, I used that whistle to break up a fight between Sailors and Marines at a recreation center, stop traffic for a group crossing a London city street, and get the attention of a crowd or individual.

Many years ago, the 42-thousand-ton *Tarawa* pulled into a foreign port for some R and R. I overheard a Sailor on the flight deck exclaim to his buddies, "I can't wait to get ashore and get shitfaced!"

I pulled out my whistle and gave it a hard and long blow.

Then, like little birdies waiting to be fed, a dozen faces stared at me on the bridge wing. I pointed to the party-ready Sailor and demanded he report to the bridge.

With nervous caution, he waited as my team completed mooring the ship to the pier. When the evolution was finished, I called the Sailor to my side. I wasn't angry, loud, or demeaning. Instead, I talked with him about choices and that his desire, above all others, to get drunk was a bad choice, emphasizing that I needed him to make better choices. I needed him to make great choices. I trusted he heard me, but as Ronald Reagan said, "Trust but verify." So, I assigned him a sober liberty buddy to ensure he behaved adequately in this foreign port. *Choices.*

That whistle on my keychain reminds me of the wonderful Sailors and Marines I had the privilege to serve with. I still occasionally pull out that whistle and slowly raise it to my lips.

My children are predictable in their response, "No, Dad. No. Don't even think about it!"

With a chuckle I return it to my pocket, unused. *Choices.*

Every day you are faced with hundreds of choices. Let me point out that if you do not feel the need to make them, someone else will make them for you.

CHOICES WE FACE – A COMPREHENSIVE AUDIT

Let's examine the everyday choices you make and how they impact your overall vision. For example, you can sleep in every day or rise and shine with enough time to prepare for the day. You can choose to eat reasonably and healthily, or you can devour that second doughnut. You can choose to play video games until 2:00am, or you can go to bed at a reasonable hour to ensure proper rest. You can choose to drink too much or imbibe in moderation.

You can choose the wrong crowd to associate with or hang out with good friends possessing honorable values. You can choose to binge movies, or you can read and improve your learning. You can choose to love improperly or love unconditionally. You can choose to provide too many examples or wrap up your point quickly.

So, please join me in choosing to lead well, lead often, and lead early. Do not fear failure but learn from it. Embrace the complexities of life and relationships. And, to do this, I want you to make great choices.

Where you start does not determine where you will end up, your choices determine your destination. When faced with decisions and choices, always choose greatness, and you will be amazed at your results and achievements. You are the sum of thousands of little choices.

Choose Greatness.

THE HIGHER YOU CLIMB THE FLAGPOLE, THE MORE YOUR BUTT HANGS OUT

You don't lose friends as you ascend to higher ranks and positions. Instead, you find out who your friends are. Leadership is scary, especially as you ascend to greater heights. Even if you do not recognize your surroundings or counterparts at different points on your journey, it is essential to know who you are at your core. Your north star is always visible when you deeply sense the legacy you wish to leave behind.

LEADERSHIP TAKEAWAYS

> ➤ Choices, small or large, shape you as an individual.Listen more than you speak.

> ➤ Making well thought out choices early on will prevent having to make a radical choice down the line.

> ➤ If you don't make a choice, others will make it for you.

ADMIRAL'S ADVICE

A ship's navigation requires choices in speed and course and helps to decide when to alter these choices. Likewise, navigating leadership and life requires multiple choices daily. As a fellow admiral summarized her leadership lessons, "Don't be a jerk! Make honorable choices; make choices of good character." Practice this early and often, and when in doubt, "Choose Greatness!"

POST SCRIPT

BURN THIS BOOK AFTER READING IT FORGET EVERYTHING YOU LEARNED IN THIS BOOK

Y ou are a leader because you read this book. Pat yourself on the back for a mission accomplished. You were curious enough to rethink your tactical approach to how you conduct your life. And quite possibly, you have learned a thing or two. I often think about the leaders in my life who were joyless or drowning in depression. I knew that I did not want to hold court with these humans. I knew I had to do things differently.

LEADERSHIP IS INCONSEQUENTIAL IF YOU DO NOT LEAD BY FOLLOWING BREADCRUMBS OF JOY.

For example, suppose you are not holding onto what brings your life a deep sense of satisfaction. In that case, your destiny will be empty. The dark space where you have sacrificed deep peace for success has the power to bring the most extraordinary men and women to their knees.

A mission-driven life plants the seeds of contentment. I fondly recite this quote often (to those who will listen):

> *"Only man can count how many seedsare in an apple, but only God can count how many apples are in a seed."*

When you lead, you plant seeds of accomplishment in those you lead, and only a greater power than you will know how many others will benefit from that leadership.

Now it is time to burn the book and forget everything you have read. There is no linear path to leadership, and your path is unique. *Sidenote: please don't burn this book, I spent a lot of time writing it. Instead, simply put it down, let this message marinate and then share it with another aspiring leader.

What I want you to know is that we are all different. The way I navigate leadership looks different than how you do it. My words are simply a framework for you to create your own roadmap. Our experiences and influences are the tapestries that make our voices unique. When we fail to be authentic, the impact we leave is minimized.

You are not me, and I am not you. You need to adapt these theories to your own life; otherwise, they will not be authentic. Even if you adopt one or two of the principles outlined in **Navigating Leadership, you'll** make a difference in your life as well as those around you. This book is meant to be a primer and to get you thinking about what you are doing right in your life and where you need to make course corrections.

After all, if I can do it, so can you.

May you enjoy leadership in a way I never imagined.
May you enjoy the sanctity of what is most
important to you.
And most important ly, may you enjoy the blessings
of the fair winds and following seas.

ANNEX 1

THREE LEVELS OF LEADERSHIP

There are three basic levels of leadership. These levels have names that suggest military-style leadership, but, in fact, they apply to any organization. Depending on the organization's size, you may be holding down multiple levels of leadership in your position.

TACTICAL

The first and basic level of leadership is the Tactical level. This is the hands-on leadership of the moment. You may be a division officer in the Navy, a squad leader in the Army, a platoon leader in the Marines, or a line-level supervisor in a manufacturing plant. The Tactical leader is concerned with the performance of her team today. Tactical leadership is also one of the highly relevant technical skills. You are the boots on the ground, in the field, at sea, or in the plant. You are running the day-to-day business of your profession. This is an exciting time as you are learning about your job and, more importantly, knowing the people that work on your team and what motivates them. You are building people skills while also honing your technical skills.

A tactical leader also identifies someone to reach out to for mentoring and advice. Some people are happy to spend their careers as tactical leaders and become the go-to expert to make things happen when an organization encounters an obstacle or roadblock.

As the Air Officer on the *USS Essex LHD-2*, aptly referred to as the Air Boss, I led a department consisting of 175 Sailors. All naval aviation professionals who worked in the ship's hangar and on the flight deck. While performing flight operations, my duty station was in the ship's control or Primary tower. From this elevated perch, I would execute the daily flight schedule, which launched and recovered our 26 US Marine helicopters and six Harrier jet aircraft.

Leadership in this environment was based on skill and longevity and not necessarily on test-taking ability. Each Sailor on the flight deck wears a color-coded deck jersey which signifies their role. Everyone on the flight deck is responsible for their safety and the safety of those on their team. Blue shirts were all the new Sailors, and after extensive training, they were assigned to a tactical leader in a yellow shirt. The yellow shirts ran the flight deck and launched and recovered aircraft. In contrast, the blue shirts did the rough and tough job of chocking and chaining aircraft under a yellow shirt's guidance. Blue shirts were the tough working crew on the deck, who, like football players, had to rest on the bench between flight cycles.

This is the ultimate scenario in tactical leadership. Due to a flight deck's loud and dangerous environment, a yellow shirt had to know their job inside and out. And they had to know their people better than anyone else and learn how to motivate and communicate with them in this volatile environment. Yellow shirts were incredible brute-force tactical leaders. They were passionate about their position, accountable to their people, and committed to the mission. The new blue shirts worked hard at their position and even harder to earn a yellow jersey. I got goosebumps watching them in action. Tactical leadership is technical and timely. It is here-and-now leadership.

OPERATIONAL

The operational leader is the glue of leadership between the tactical and the strategic. Therefore, the operational leader must understand and carry out strategic guidance and be conscious of their team's structure and current leadership. The operational leader is, more often than not, a leader of leaders.

Too often, individuals in an operational leadership position lead tactically. This may be considered micromanagement, whether intentional or unintentional. Some people in higher leadership positions will do the things they are most familiar with or comfortable with. Being tactically proficient leads to more responsibility and growth. If you do not understand your new leadership role, you may fall into your previous role. I've watched great ship handlers become captains and then continue to handle the ship at all times. This results in disillusionment and discontent in those learning their own tactical skills as ship handlers. Operational leaders must empower their subordinates for growth.

In NATO Northwood, we were NATO's northern maritime component command. Once a year, we would conduct a major exercise at sea. Northwood acted as the operational command center directing the operation.

During a portion of the exercise, we were hunting an adversary submarine. Our headquarters chief of staff had been a successful Dutch destroyer captain and well-versed in submarine hunting. From the command center hundreds of kilometers away from the action, he gave precise directions on where the friendly destroyer should sail, what electronic devices to employ, and where to set acoustical boundaries.

I jumped in before the orders flew and told the Chief of Staff that he was an operational commander acting like a tactical operator. I assured him the destroyer captain on the scene only needed to be ordered to prosecute the submarine. She would employ their weapons and sensors to the best of her ability. The elite leader understands that the skills that got them promoted are not necessarily the skills they need in their new role. Those skills will be valuable, but a new level of leadership is needed in this new role.

Operational leaders lead leaders. If you are doing a subordinate's job, who is doing your job? Operational leaders empower their subordinates to perform at their best and provide the resources to do so.

STRATEGIC

The strategic level of leadership is concerned with creating a compelling vision in alignment with the mission of the organization and attracting others to that vision." The leader's strategic role is to envision the future and a path to an organization's future. Finally, once a compelling vision for the future is developed, the strategic leader focuses on the resources needed and roadblocks that require clearing as the organization progresses.

Strategic leaders are visionaries, and their compelling vision inspires others at all levels of leadership. When I work with military leaders and entrepreneurs on strategic issues, I have a fundamental question for their organization. "As an organization, who are you?"

Answering this usually involves a great deal of stumbling around as they try to articulate a response. When the realization sets in that they haven't put that much thought into this question, I give them an assignment. While this assignment can work on many levels, it should be performed at the strategic level.

I ask them to develop a document titled, *Who We Are.* This two-to-three-page document makes a tremendous difference in leading an organization. If you know who you are, then you also know who you are not. This also is an excellent tool when counseling subordinates concerning personal and organizational performance.

Tactical, Operational, and Strategic

Knowing where you fit in these three levels of leadership will enhance your performance and accelerate your growth. In addition, understanding your position at the leadership level will prevent subordinates' frustration, thus ensuring everyone in an organization is operating in their role for the benefit of others and your organization's mission.

ANNEX 2

DETAILS ABOUT THE AUTHOR

Following service as Executive Director of the Association of the US Navy, Rear Admiral Garry Hall received a political appointment to serve as a Senior Director on the National Security Council (NSC) and Special Assistant to the President for National Security Affairs. At the NSC, he was responsible for Human Rights, Humanitarian Assistance, Immigration and Migration, Atrocity Prevention, United Nations Operations, Democracies, and Fragile States. He supported these portfolios and led a team of career professionals dedicated to each of these functional areas.

A United States Naval Academy graduate, he served 35 years on active duty. As a naval aviator, he flew the venerable *SH-2F Light Airborne Multi-Purpose (LAMPS Mark I)* anti-submarine warfare helicopter during his early career. In addition, he flew operationally in Helicopter Anti-Submarine Squadron Light Thirty-Seven (HSL-37.) While in HSL-37, based in Hawaii, he deployed for seven months in the *USS Rathburne FF 1057* and eight months in the *USS Cushing DD-985*.

Following HSL-37, the Admiral was selected as an instructor pilot in the SH-2F in HSL-31, the Fleet Replacement Squadron in San Diego, California. There he prepared pilots and aircrew for their assignments in the fleet.

The Admiral was again ordered to an operational fleet squadron, HSL-35, in San Diego. He deployed in *USS Reasoner FF-1063* for seven months as Officer in Charge (OINC) of a detachment. Following that deployment, he was honored as the Navy's Helicopter Pilot of the Year for Sustained Performance. He finished that tour as the Operations department head.

Hall was then competitively selected as the Aide de camp to the Vice Director of Joint Strategic Target Planning Staff (JSTPS) on Offutt Air Force Base in Omaha, Nebraska. There he was the personal aide to two consecutive three-star admirals. While on this tour, he was promoted to Commander one year early and selected to command an operational squadron in Atsugi, Japan.

Admiral Hall was the first commanding officer of an SH-60B helicopter squadron in Japan, the HSL-51 Warlords. Using the vision and goal-setting methods described in this book, he led the squadron from formation and establishment to a combat-ready and award-winning squadron in its first year of eligibility. In less than two years, HSL-51 earned the Battle Efficiency award as the most combat-ready SH-60B squadron in the Pacific Fleet. In addition, Hall was ranked number 1 among 26 squadron commanders.

Leaving Japan, Hall reported to the USS Essex LHD-2 as the Air Department Head, commonly referred to as the Air Boss. During this tour, he led the Air Department with 175 Aviation professionals. In addition, he conducted air operations for ten months at sea in the Pacific and Indian Oceans. While on deployment, Hall was selected for bonus command of the Fleet Replacement Squadron HSL-41 to train fleet-bound pilots, aircrew, and maintainers in the SH-60B aircraft.

Following his second highly successful squadron command, Hall was selected to command the capital ship *USS Tarawa LHA 1,* which had a complement of 1000 Sailors, 2000 Marines, 32 aircraft, and 3 landing craft. On *Tarawa*, he served as the Executive Officer for eighteen months, followed by taking command of the ship for another eighteen months. During this tour, he deployed twice for six months each. In addition, he spent six months in the shipyards of Bremerton, Washington, away from his homeport.

Returning from his deployment on *Tarawa*, he reported to Commander Naval Air Forces as the Admiral's executive assistant. September 11th occurred shortly into this assignment, and Captain Hall was selected as a new Rear Admiral or One-star. His stars were pinned on, and he left San Diego to support General Tommy Franks as his Information Operations Officer at the beginning of Operation Iraqi Freedom. Next, Hall was reunited with Darlene and his family, serving with NATO in the United Kingdom. There he learned the ins and outs of NATO and honed his diplomatic skills during the early years of the Iraq war.

Following NATO, he was ordered back into the operational Navy. He commanded an Expeditionary Strike Group deployed to the Middle East. He was responsible for the amphibious forces in the Middle East. In addition, he led various tasks, from protecting the Iraqi oil infrastructure at sea and fighting piracy off the coast of Africa.

The Admiral's final uniformed assignment was to lead a Senior Service College at National Defense University and academically prepare future leaders of our military in national security and resource management.

Following active duty, Hall has worked in the defense industry and in the non-profit association industry. He has also brought his leadership and risk management skills to the Catholic church serving as the Chairman of the Archdiocese of the Military Services Review Board for the protection of children. He also served a four-year term on the

National Review Board advising the United States Conference of Catholic Bishops on child protection.

Admiral Hall holds a B.S. in Marine Engineering from the U.S. Naval Academy, an MBA from Southern Illinois University at Edwardsville, and is a graduate of National Defense University's Capstone Course.

Hall was recognized throughout his career with unit and personal awards, each recognition he attributes to the hard work of his Sailors and Marines. In May of 2023, his high school, Kenmore West High School in Kenmore New York, inducted Hall into their Corridor of Honor in recognition of those who "throughout their lives have exhibited those values, qualities, and characteristics traditionally held and encouraged by Kenmore West High School."

Admiral Hall now resides in Hill Country, Texas, with the love of his life Darlene, enjoying the good life. Darlene continues to be his secret weapon to success. She is the family's anchor, leading up and down the family tree of four generations. Hall continues to speak publicly on leadership, faith, and business and hosts the popular podcast, **The Admiral's Almanac**.

And above all, Hall is a super-duper advocate for young leaders changing the world. And Hall does this with his irreverent wit and wisdom.

Career Timeline

1976
Graduate from the U.S. Naval Academy, Annapolis MD. Rank of Ensign

1977
Met and Married Darlene Allred, Corpus Christi Texas.

1978
Earned "Wings of Gold" Pensacola FL

1981-1984
Instructor Pilot with HSL-31 Coronado CA. Flying the SH-2F helicopter. Lieutenant.

1979 - 1981
Stationed in Hawaii with Helicopter anti-submarine Squadron Light Thirty-Seven, HSL-37 deploying in USS Rathburne and USS Cushing. Flying the SH-2F helicopter. Rank of Lieutenant Junior Grade and promoting to Lieutenant.

1985-1987
Lieutenant Commander assigned to HSL-35 Deploying in USS Reasoner. Selected as Helicopter Pilot of the Year 1986, flying the SH-2F helicopter.

1987-1990
Aide de camp to Vice Director Strategic Targeting, Omaha NE. Selected early for Commander and selected to Squadron Command.

1994-1995
Air Boss on the Assault Ship USS Essex. LHD-2, deployed overseas. selected for second squadron command. Rank of Commander.

1991-1993
Commanding Officer HSL-51 in Atsugi Japan. Squadron recognized for Battle Efficiency. Flying the SH-60B helicopter. Rank of Commander.

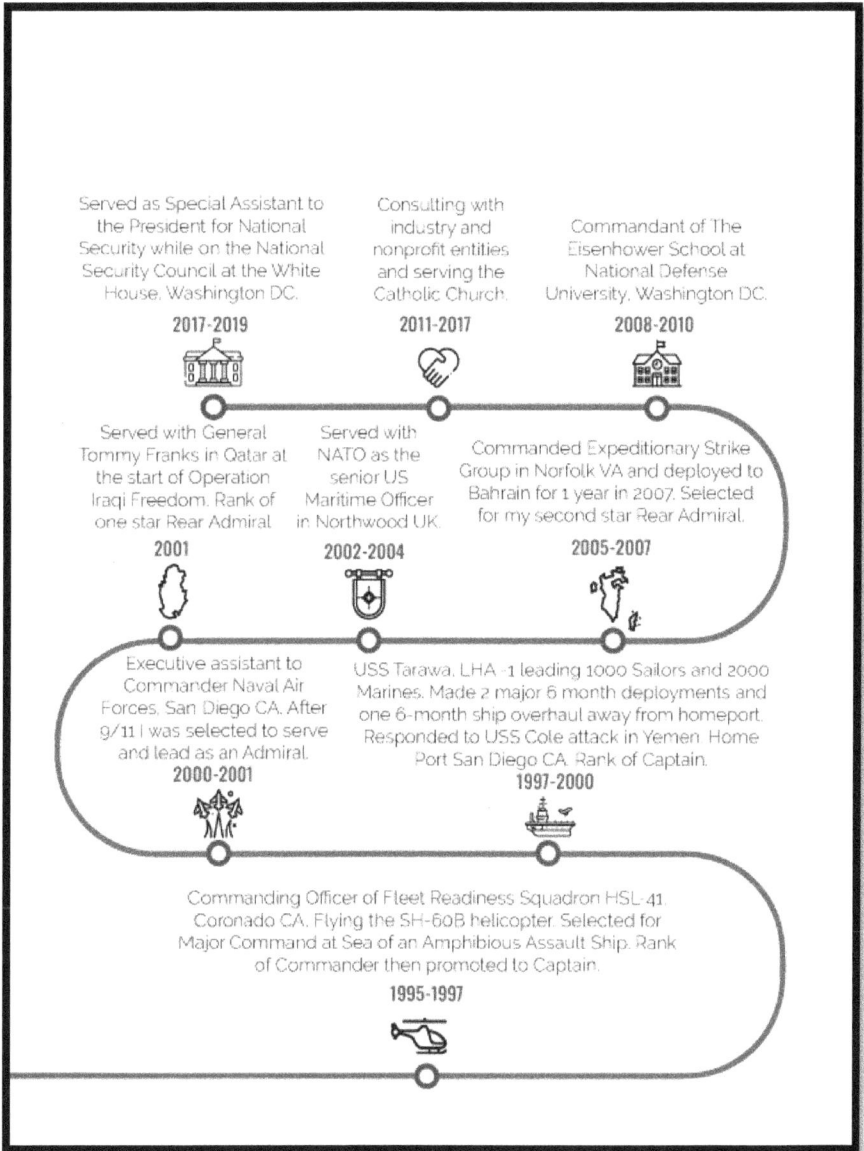

Served as Special Assistant to the President for National Security while on the National Security Council at the White House, Washington DC.

2017-2019

Consulting with industry and nonprofit entities and serving the Catholic Church.

2011-2017

Commandant of The Eisenhower School at National Defense University, Washington DC.

2008-2010

Served with General Tommy Franks in Qatar at the start of Operation Iraqi Freedom. Rank of one star Rear Admiral.

2001

Served with NATO as the senior US Maritime Officer in Northwood UK.

2002-2004

Commanded Expeditionary Strike Group in Norfolk VA and deployed to Bahrain for 1 year in 2007. Selected for my second star Rear Admiral.

2005-2007

Executive assistant to Commander Naval Air Forces, San Diego CA. After 9/11 I was selected to serve and lead as an Admiral.

2000-2001

USS Tarawa, LHA -1 leading 1000 Sailors and 2000 Marines. Made 2 major 6 month deployments and one 6-month ship overhaul away from homeport. Responded to USS Cole attack in Yemen. Home Port San Diego CA. Rank of Captain.

1997-2000

Commanding Officer of Fleet Readiness Squadron HSL-41. Coronado CA. Flying the SH-60B helicopter. Selected for Major Command at Sea of an Amphibious Assault Ship. Rank of Commander then promoted to Captain.

1995-1997

ANNEX 3

A RECOMMENDED READING LIST

There are many recommended reading lists produced by influential business and military leaders. A quick internet search will produce them. I only add a few books to augment any one of those lists.

- ***The Effective Executive: The Definitive Guide to Getting the Right Things Done*** *by* Peter Drucker. This book has been given to all new generals and admirals over many years by former Speaker of the House, Newt Gingrich. It is a great way to think about how you manage your time as an executive or military leader.
- ***Leading Change*** by John P. Kotter. I like the process Kotter outlines on leading change. When you lead, often you are leading change. This book will make you think on creating a succinct compelling idea to motivate your teams.
- ***The SPEED of TRUST: The One Thing That Changes Everything*** by Stephen M. R. Covey. Understanding the power of trust in leading is a must. Covey shows you how to build trust and how to recover trust if you have lost it with a team.
- ***The Leader's Bookshelf*** by Admiral James Stavridis. This author is a friend and classmate. He is also a prolific writer. In this work he has provided the thoughts of many leaders and the books they read that have influenced their leadership. You can

start anywhere in this book and read as much or as little as you want in one sitting and make a difference in your leadership.

- ***Think and Grow Rich*** by Napoleon Hill. This almost a century old book, may have some outdated language but it continues to have a powerful and positive effect on its readers. I share it often and always to positive reviews.

THANK YOU

I learned early in my career that the more people you thank in a speech the more others feel left out. With that in mind, there are many I must thank.

I thank every member of the Bicentennial Class of 1976 from the U.S. Naval Academy. Bound by four years where Severn meets the tide, we were forged as leaders and friends for life. Specifically, I want to thank two of my roommates:

Colonel P. R. "Hutch" Hutcherson, USMC (ret.) Hutch has been my peer mentor for 50 years dispensing straightforward advice and counsel and setting the example of unwavering moral courage. He was the Best Man in my wedding, and we continue to talk weekly.

Captain David "Sweens" Sweeney, Northwest Airlines (ret.) Dave was my roommate in Annapolis and for most of our flight training. We were partners "in crime" at the academy and during flight school. We amazed our flight school classmates by graduating at the top of our flight class.

Thank you to my two helicopter community mentors that saw potential in a young officer and developed and promoted that potential.

Captain Ken "Cannonball" Marion, US Navy (ret) you showed us all class and professionalism in the air and on the ground. May you rest in peace until we are on the flight schedule together again.

Captain Garnett "Sandy" Clark, US Navy (ret) your lifetime of council and friendship helped me navigate faith, family, and profession.

Thank you to my Spiritual advisors who saw something in me needing direction. Thank you Navy Chaplain Don denDulk, Captain US Navy (ret) may you rest in peace. You taught me unconditional love as a father.

Thank you, Bishop William Byrne, friend, mentor, and leader of the Catholic Faith, you brought me closer to my faith. You are a joyful leader.

Thank you to two of my senior enlisted advisors and leaders. Thank you, Master Chief Rod MacAfee, US Navy, my department Master Chief while I was Air Boss on *USS Essex*, you showed amazing deck plate leadership and prepared me for major command of *USS Tarawa*.

Thank you, Master Chief Jerry Haueter, US Navy (ret.) You are the consummate leader of Sailors and Marines. Your personal attention to my safety and well-being while also leading a thousand sailors deployed to the Middle East was an amazing experience.

Thank you to my many copilots in the air and on the ground, some memorable ones "Race" Bannon, "Kid" Lumme, "Mookie" Wilson, "Madonna" McDonough, "Cuervo" Cavazos, Jim callsign "Jim" Cox, "Magnum" Key, and Bill callsign "Bill" Stedman.

Thank you to the Executive Officers that backed me up and advised me well during my squadron commands. Captain Dave Landon, US Navy (ret) with the Warlords of HSL 51 in Atsugi Japan, Captain Marty Keeney, US Navy (ret,) and Captain Steve Greene, US Navy (ret) with the Seahawks of HSL 41 in San Diego California, and Captain Al Nugent, US Navy (ret) with *USS Tarawa, LHA-1.*

Thank you to all my department heads, executive officers, and commanding officers who overlooked my weaknesses and developed my strengths as a leader.

Thank you, Vice Admiral Ron Eytchison, US Navy (ret) who I had the honor of being his Aide de Camp. You showed me by your example how to be an ethical and humble leader and husband as an admiral.

Thank you, Admiral John Nathman, US Navy (ret) who I had the honor of being his Executive Assistant. You showed me by your example how to be a strategic leader and develop vision for a fleet. You prepared me for serving as a Rear Admiral.

Thank you to my 3 Aides that supported me as a Flag Officer.

Commander Mike Lee, US Navy, Captain Brett Elko, US Navy, and Captain John Anderson, US Navy. Each one an ethical and humble leader making a difference every day in the Navy.

Finally, no book comes to life without the help of writing, editing, publishing, and promoting professionals. Thank you, Julie Lokun, Jessica Dalby, Maccabee Griffin, Renee Wertz, Kelsey Pitarra and Heidi Dunstan for bringing this book to life.

Thank you for reading, Navigating Leadership,
Making a PACT with Excellence.

If you would like Garry Hall to speak at your event,
contact Dan LaBert at FWA@FlagOfficers.us

To hear more from Admiral Hall

Listen to his podcast "The Admiral's Almanac"
wherever you get your podcasts.

Or www.admiralsalmanac.com

If you enjoyed this book please leave a review on Amazon.

Until we meet once more, here's wishing you a happy voyage home!

www.ingramcontent.com/pod-product-compliance
Lightning Source LLC
Chambersburg PA
CBHW040923210326
41597CB00030B/5159